T0176656

Statistics

A Concise Mathematical Introduction for Students, Scientists, and Engineers

David W. Scott
Rice University
Houston, Texas

Registered Offices
John Wiley & Sons, Inc., 111 River Street, Hoboken, NJ 07030, USA
John Wiley & Sons Ltd, The Atrium, Southern Gate, Chichester, West Sussex, PO19 8SQ, UK

Editorial Office
9600 Garsington Road, Oxford, OX4 2DQ, UK

For details of our global editorial offices, customer services, and more information about Wiley products visit us at www.wiley.com.

Wiley also publishes its books in a variety of electronic formats and by print-on-demand. Some content that appears in standard print versions of this book may not be available in other formats.

Library of Congress Cataloging-in-Publication Data has been applied for

Paperback ISBN: 9781119675846

Cover Design: Wiley
Cover Image: Courtesy of David W. Scott

Set in 9.5/12.5pt STIXTwoText by SPi Global, Chennai, India

Printed and bound by Quad/Graphics

VACAEEBE5-3511-4717-97D4-4F059DF7B6AE_062320

To my parents, John and Nancy Scott

Contents

Preface

My aim in writing this book is to provide a self-contained, one-semester probability and statistics introduction that covers core material without ballooning into a huge tome. Since statistics requires an understanding of distributions and relationships (for example, predicting y from x), some introductory knowledge of multivariate calculus and linear algebra will be assumed. Examples will use the **R** language, but they can easily be modified to other systems such as Matlab. Mathematica will be used for symbolic computations. JMP can be used to perform statistical tests in a unified manner.

The course divides naturally into three sections: (1) classical probability; (2) distribution functions, density functions, and random variables; and (3) statistical inference and hypothesis testing.

In selecting material to include, I have favored models that follow directly from simple, intuitive assumptions. I have also favored statistical topics that are widely used. In this era of data science, I have occasionally selected new topics that are relevant and easily understood. For example, robustness is relevant because bad data or outliers can adversely affect classical methodology.

Students who have taken AP Statistics will have an advantage in that they will have seen a large number of cookbook statistical procedures and tests. We will cover only a selection, as the mathematical foundations (or outline thereof) will be of equal interest here. Often we will sacrifice mathematical rigor in favor of an engineering-level understanding without apology. Motivated students will naturally follow this course with more mathematically rigorous courses in statistics, probability, and stochastic processes. Reading about other statistical tests and methods should be straightforward after mastering the material covered here.

I have included a handful of problems and case studies, to keep things simple. There will be a live course website with numerous sample problems and exams. Instructors with special interests can easily insert their own examples and problems in appropriate sections. The URL for the additional course material is

http://www.stat.rice.edu/~scottdw/wiley-dws-2020/

The directory contains problems, sample exams, and the pdf file all-figs.pdf, which displays all 57 figures, including 45 color diagrams. The author may be reached at scottdw@rice.edu

I wish to thank James R. Thompson, who introduced me to the beauty of model building and statistical thinking. He served as one of my thesis advisers, directing me into the joys of nonparametric modeling. He was in turn highly influenced by his thesis adviser, John W. Tukey, one of the most important statisticians of the 20th century. Tukey's contributions ranged from the fast Fourier transform to the body of graphical work introduced in his monograph *Exploratory Data Analysis*. Their ideas appear throughout this book.

Houston, Texas   *David W. Scott*
September, 2019

1

Data Analysis and Understanding

The field of statistics has a rich history that has become tightly integrated into the emerging field of data sciences. Collaboration with computer scientists, numerical analysts, and decision makers characterizes the field. The role of statistics and statisticians is to find actionable information in a noisy collection of data. Every field of academic endeavor encounters this problem: from the electrical engineer trying to find a signal in a noisy channel to an English professor trying to determine the authorship of a contested newly discovered manuscript.

There are two basic tasks for the statistician. First is to characterize the distribution of possible outcomes using a batch of representative data. An actuary may be asked to find a dollar loss for car accidents that is not exceeded 99.999% of the time. An economist may be asked to provide useful summaries of a collection of income data. The histogram is our primary tool here, an idea that did not appear until the 17th century; see Graunt (1662), who analyzed death records during height of the plague outbreak in Europe.

The second task is that of prediction. A bank may wish to understand how credit risk is related to other information that may be available. A mechanical engineer may wish to understand the risk inherent in a new design under extreme conditions. Methods for performing this task underlie many algorithms today, for example, translating foreign languages or image recognition.

The mathematical backbone of all of our statistical methods is probability theory. Thus we study the basics of probability theory and random variables in the first part of this course. Statistical methods and the basics of statistical decision theory form the core of the middle third of this course. Specific tests and data analysis approaches finish our study.

1.1 Exploring the Distribution of Data

Tukey (1977) introduced a number of data summaries in his book *Exploratory Data Analysis*. Many are based on quantiles or percentiles of the data vector. Percentiles are particular choices of the sorted data. The middlemost is the **median**, or the 50th percentile. As a measure of spread, Tukey focused on the distance from the 25th to the 75th percentiles, the so-called *interquartile range* (IQR). A three-point summary would list these percentiles. Instead Tukey popularized the box-and-whiskers plot, which is a five-point summary. The additional two points are intended to capture 99% of the data.

Statistics: A Concise Mathematical Introduction for Students, Scientists, and Engineers, First Edition. David W. Scott.
© 2020 John Wiley & Sons Ltd. Published 2020 by John Wiley & Sons Ltd.

These are drawn at a distance of $1.5 \times$ IQR from the two quartiles. Any points outside these whiskers are plotted as potential *outliers*.

1.1.1 Pearson's Father–Son Height Data

We illustrate these ideas on a set of data collected by Karl Pearson over a century ago. He recorded the heights of $n = 1{,}078$ fathers and an adult son. In the left frame in Figure 1.1, we display a box-and-whiskers plot of these data. We see that the sons are taller than their fathers by about an inch. There are also more potential outliers among the sons for some reason.

In the middle frame of Figure 1.1, we show Tukey's stem-and-leaf plot of the 1078 *differences* of the heights of each son and his father. The range of the data is $(-8.96, 11.24)$ and the first seven sorted values rounded to one decimal place are $(-9.0, -7.9, -7.4, -7.3, -7.1, -6.6, -6.4, \dots)$. Each data point is decomposed into a stem and a leaf digit. Thus -9.0 has a stem of -9 and a leaf of 0. The top line is actually $\boxed{-9 \mid 0}$, although it is too small to see. With so much data, each stem is broken into two lines to provide more detail. Thus the next two lines show a stem of -8 but no leaves $\boxed{-8 \mid}$ twice. The fourth line shows $\boxed{-7 \mid 9}$ and the fifth line reads $\boxed{-7 \mid 431}$ and so on. This figure was generated using the **R** command $\boxed{\texttt{stem(x, scale=3)}}$; **R** Core Team (2018). (The default `scale=1` has half as many stems.) Thus the stem-and-leaf plot shows the frequency count of points for each stem as character strings.

In the right frame of Figure 1.1, we show the frequency counts in a **histogram**. The histogram uses a parameter h called the bin width to construct an equally spaced mesh $\{\dots, -2h, -h, 0, h, 2h, \dots\}$. Then we count the number of points in each interval. These counts are displayed as a bar chart. (The histogram can use any anchor point, although 0 is a common choice.) For the histogram shown, the anchor point selected was 0, and h was chosen using Scott's rule $h = 0.945$; see Scott (1979). This rule is discussed in Section 9.1.4.1. The default choice in **R** function `hist` is Sturges' rule, discussed in Section 9.1.4.3, which chooses 11 bins with $h = 2$ (not shown).

The choice of h is often considered a matter of convenience. The stem-and-leaf plot using one-digit integer stems limits its choices. By way of contrast, any positive real number h

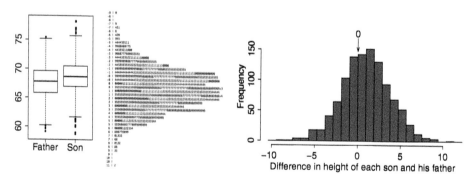

Figure 1.1 Displays of the father–son height data collected by Karl Pearson: (left) box-and-whiskers plot; (middle) stem-and leaf plot; (right) histogram.

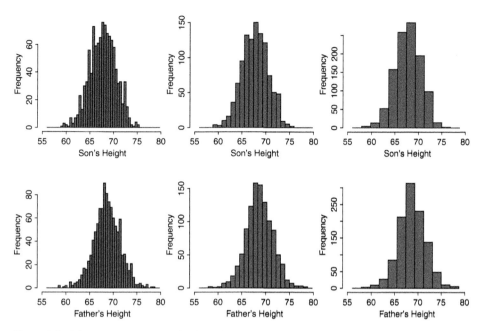

Figure 1.2 Histograms of the sons' heights (top row) and fathers' heights (bottom row) using three bin widths: $h/2$, h, $2h$ from left to right; see text.

can be used in a histogram. In Figure 1.2, we show the histograms using h by Scott's rule, as well as $h/2$ and $2h$. Loosely speaking, the histograms using $2h$ are missing useful information, while the histograms using $h/2$ display spurious detail. We discuss strategies for finding the best choice of h in Section 9.1. In any case, the histogram is a powerful tool for understanding the full distribution of data.

1.1.2 Lord Rayleigh's Data

In *Exploratory Data Analysis*, Tukey (1977) demonstrates the box-and-whiskers plot using the Lord Rayleigh data, which measure the weight of nitrogen gas obtained by various means; see Table 1.1. Discrepancies in the results led to his discovery of the element argon. Rayleigh made $n = 24$ measurements from 1892 to 1894, with a mean of 2.30584

Table 1.1 Lord Rayleigh's 24 measurements (sorted) of the weight of a sample of nitrogen. The first 10 came from chemical samples, while the last 14 came from pure air.

2.29816	2.29849	2.29869	2.29889	2.29890
2.29940	2.30054	2.30074	2.30143	2.30182
2.30956	2.30986	2.31001	2.31010	2.31010
2.31012	2.31017	2.31024	2.31024	2.31026
2.31027	2.31028	2.31035	2.31163	

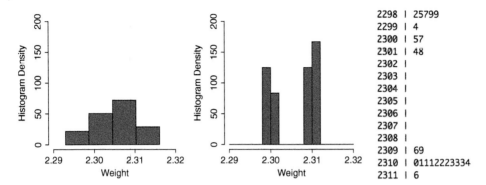

Figure 1.3 Displays of Lord Rayleigh's 24 measurements of the atomic weight of nitrogen gas. (Left) Histogram with four bins; (middle) a second histogram; (right) stem-and-leaf display using the R command `stem(rayleigh,scale=4)`.

and a standard deviation of 0.00537. It is common to assume such measurements of a fundamental quantity are normally distributed. Multiple experiments are run and the results averaged in the presumption that a more accurate estimate will result.

In the left frame of Figure 1.3, we display a histogram with four (carefully selected) bins. The histogram is shown on a density scale, rather than a frequency scale, so that the area of the shaded region is 1. We shall see in Problem 1 that this is accomplished by dividing the bin counts by nh.

The first histogram in Figure 1.3 hides the interesting structure contained in the small dataset. The second histogram and stem-and-leaf plot show the two clusters quite clearly. Charting of data before the 1900s was not common, and looking at a table of the data would typically not reveal this feature. It turned out that Lord Rayleigh had combined various sources of the gas with several purifying agents and extraction methods. The samples originating from "pure air" were "contaminated" with argon. For the discovery of argon, Lord Rayleigh was awarded the Nobel Prize in Physics in 1904.

1.1.3 Discussion

Finding structure in data is a primary goal of data science. Graphical methods are powerful approaches to discovering unexpected or hidden structure. Some of these methods are better suited to small datasets. In a multivariate statistics course, we will learn how to analyze data with more than one variable. Modern genetic datasets often result in more than 20,000 variables!

1.2 Exploring Prediction Using Data

The second fundamental task of statistics is prediction. Data for this task are typically ordered pairs, $\{(x_1, y_1), (x_2, y_2), \dots, (x_n, y_n)\}$. The goal is to predict the value of the y variable using the corresponding value of the x variable. For example, we might try to predict a son's height (y) knowing the father's height (x). Or a bank contemplating a mortgage loan may use a person's credit score to predict the probability the person will default on the loan.

The initial step is to plot a **scatter diagram** of the n data points in order to determine if there is a strong relationship between x and y. The relationship, if it exists, is linear or nonlinear. If knowledge of x does not convey any information about the value of y, then the scatter diagram will have no slope or trends, with y values just scattered around their average.

1.2.1 Body and Brain Weights of Land Mammals

In the left frame of Figure 1.4, we plot the brain and body weights of 62 land mammals from the **R** MASS library. The relationship, if any, is hard to discern since 59 of the measurements are overplotted near the origin. We might choose to exclude the two elephant measurements and even the human data point as **outliers**, and then replot.

However, Tukey introduced a power transformation ladder to re-express a variable x:

$$\cdots \quad x^{-2} \quad x^{-1.5} \quad x^{-1} \quad x^{-1/2} \quad \log(x) \quad x^{1/2} \quad x \quad x^{1.5} \quad x^2 \quad \cdots \ ;$$

see Problem 3 for an explanation of why $\log(x)$ is used in place of $x^0 = 1$ when $\lambda = 0$.

In the right frame of Figure 1.4, we use the log function to dramatic effect. There clearly is a strong relationship that allows highly accurate prediction of the log(brain weight) of a land mammal knowing its log(body weight). (The body weight is easily measured for a living specimen, but not its brain weight.) Moreover, the relationship appears to be linear. In this re-expressed scatter diagram, the two or three outliers identified in the first plot are no longer outliers.

1.2.2 Space Shuttle Flight 25

The 25th launch in the Space Shuttle program was scheduled for 22 January 1986, but postponed for various reasons each day until 28 January. The temperature had dropped to 28° overnight, and it was 36° when the launch was attempted at 11:38 a.m. During the first 90 s,

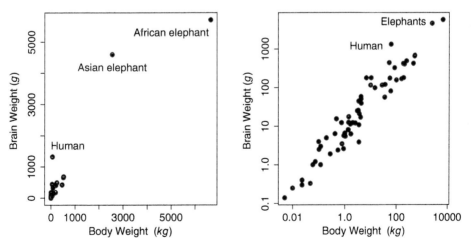

Figure 1.4 Scatter diagrams of the raw and log-transformed body and brain weights of 62 land mammals.

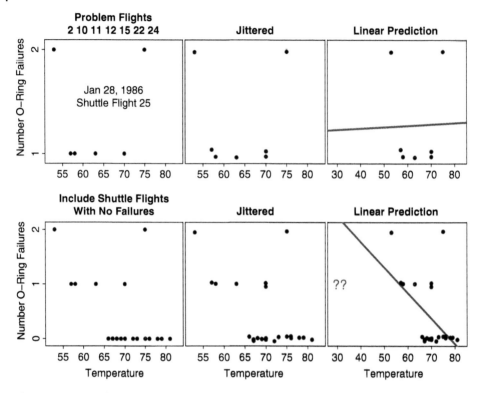

Figure 1.5 Analysis of the number of O-ring failures for the first 24 Space Shuttle launches; see text.

several O-rings on the solid rocket boosters failed, leading to a catastrophic explosion and loss of all seven crew members. Scientists knew previous shuttle flights had occasionally experienced one or two O-ring failures, but a launch had never been attempted at freezing temperatures. Varying opinions of the safety were provided to the launch director, who eventually decided to proceed. One of the data analyses is reproduced in the first row of Figure 1.5.

In the heading of the scatter diagram in first frame, we see a list of the 7 (of the first 24) shuttle flights that experienced 1 or 2 O-ring failures. Two failures were observed at the lowest temperature of 53°, which was well above the temperature range of 28–36° on the day of the disaster. Strangely, two failures had also been observed at the highest temperature of 75°.

In the second frame, we have **jittered** the data by adding a little uniform noise. This reveals that there were two data points superimposed at $(70, 1)$; jittering broke that tie. In the third frame, the data are replotted, but with an expanded x-axis to include 28°. Would you have supported the decision to launch? A least-squares line (discussed in Chapter 8.5) is superimposed. This line suggests that, if anything, lower temperatures might result in fewer O-ring failures. Thus the launch was attempted.

However, in a re-analysis of these data, we have included the shuttle flights that experienced no O-ring failures. Now the final frame suggests that two or more O-ring failures are quite likely at 28–36°. The question of including or excluding data is a difficult problem in practice. In other settings, including non-event data can bias the analysis in the wrong direction. As we saw in the brain-body weight data, excluding the two or three outliers was not necessary. However, in Rayleigh's nitrogen data, excluding an entire cluster of outliers as *bad data* would have postponed the discovery of argon.

1.2.3 Pearson's Father–Son Height Data Revisited

We have explored the two variables in this dataset individually, but there is an obvious question of how accurately a son's height can be predicted knowing his father's height. In the first frame of Figure 1.6, we display a scatter diagram of the $n = 1078$ pairs. This diagram clearly shows a positive tilt, consistent with the expectation that the sons of tall fathers are tall, and vice versa; however, the strength of the relationship does not seem as strong as in the brain–body weight dataset.

In the top right frame, we have placed a red dot at the location of the average heights of the fathers and sons. We have also drawn a straight line fit using the intuitive equation $y = x$. However, the equation $y = x + 1$ is an improvement, since we observed earlier that sons were 1 inch taller than their fathers on average. As a reference, we have also included a horizontal line at the average heights of the sons. This line would be appropriate if there were no information about a son's height to be gleaned from his father's height; but a positive relationship (correlation) is clear.

Galton (1886) was one of the first to observe that many scatter diagrams observed in nature have an appearance similar to that in Figure 1.6. He noted that the shape appeared

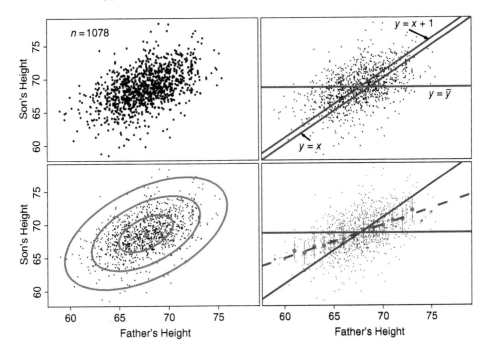

Figure 1.6 Father–son height data collected by Karl Pearson.

elliptical, so he superimposed elliptical contours over the scatter diagram. The bottom left frame in Figure 1.6 shows three (nested) ellipses for these data. Recall that a general ellipse has five parameters: two for the center of the ellipse; two for the horizontal and vertical scales; and a fifth called the *eccentricity*. Galton focused on this fifth parameter, and the **correlation coefficient** was the result. Ironically, this parameter is often referred to today as **Pearson's correlation coefficient**.

In the final frame, we take advantage of the large sample size to try to understand if the prediction (as weak as it may be) might be linear or nonlinear. For integer values of the rounded fathers' heights, we compute a three-point summary of the corresponding sons' heights. The red dots are the arithmetic average of the sons' heights. The vertical lines display the (conditional) interquartile range. The final two red dots on each end are based on only a few points, so that the IQR can not be computed. These four red dots are shown in a smaller font size to indicate that even the averages are not so reliable.

We see that these summary points clearly suggest a linear rather than a nonlinear fit. We also see that the two blue reference lines from the second frame, namely $y = x + 1$ and $y = \bar{y}$, both miss badly. A new (dashed) line with slope of $1/2$ appears to capture the linear trend quite well. The relationship between this slope and the correlation coefficient, as well as a genetic explanation, will be discussed in Chapter 4.1.5.

1.2.4 Discussion

These rather substantial examples illustrate the search for structure in distribution and prediction problems, as well as practical problems and cures that may be encountered. A more formal statistical approach to these questions will be introduced in the third part of this course. Probability theory will be the theoretical basis for many of these models, so we make it the focus of the next few chapters.

Problems

1.1 A **frequency histogram** of continuous data is constructed by counting the number of data points that fall into equally spaced bins of width h. h is called the **bin width**. Typically the bin edges are $0, \pm h, \pm 2h, \pm 3h$, and so on. If the bin count in the kth bin is denoted by v_k, then the frequency histogram is defined as

$$\boxed{\hat{f}(x) = v_k, \qquad \text{for } x \text{ in the } k\text{th bin.}} \tag{1.1}$$

(a) Show that the total area of the frequency histogram is nh, where $n = \sum_k v_k$.
    Hint: the histogram is made up of rectangular blocks of width h and height v_k.
(b) A **probability histogram** is defined to have total area of one. Show that the following definition of a histogram has area equal to one:

$$\boxed{\hat{f}(x) = \frac{v_k}{nh}, \qquad \text{for } x \text{ in the } k\text{th bin.}} \tag{1.2}$$

1.2 One of the most famous epidemiological cases occurred in 1854 when Dr. John Snow successfully tracked down the source of an outbreak of **cholera** in the London suburb of SoHo. He mapped the households of some 500 victims over a 10-day period that lived within a quarter of a mile of each other. However, many tens of thousands had died of cholera in England during the prior two decades. Dr. Snow believed contaminated water was a primary cause. Just as in the Space Shuttle example, there are choices of an appropriate time interval and the geographical extent that can influence our conclusions. Using the descriptions and maps conveniently assembled at
http://www.ph.ucla.edu/epi/snow/snowcricketarticle.html,

discuss the evidence and choices that were and could have been made. *Hint: these data have been conveniently collected in CRAN Library* HistData *by Friendly (2018). Look at the help file for dataset* snow *and its example code.*

1.3 (a) The Tukey power transformation of a variable x is x^λ for any non-zero $\lambda \in \mathbb{R}^1$. To better understand why the $\log(x)$ is used in place of x^0 when $\lambda = 0$, we consider the linear re-expression of the Tukey transformation given by the formula (Box and Cox (1964))

$$\frac{x^\lambda - 1}{\lambda}. \tag{1.3}$$

Since (1.3) is 0/0 when $\lambda = 0$, use l'Hôpital's rule to find the limit transformation as $\lambda \to 0$. The scatter diagram using either formula for fixed non-zero λ will be visually identical. Formula (1.3) is referred to as the Box–Cox transformation; see Figure 1.7.

(b) Sometimes the transformation $\log(1 + x)$ is used in place of $\log(x)$ when $x \geq 0$ and x can take on the value 0. In this case, the original and transformed values of 0 are both 0. Try this form on the body–brain data and compare to Figure 1.4.

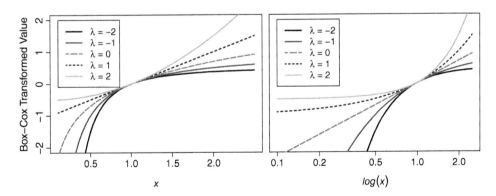

Figure 1.7 Box–Cox transformation on natural and log scales.

2

Classical Probability

Early developments in probability may be attributed to gamblers who reached out to scientists to better understand games of chance. For simple games of dice or cards with equally likely outcomes, probabilities could be easily calculated. However, for more complicated games, gamblers often used intuitive approximations that they knew were at odds with their extensive experience. Progress was made by scientists such as Pascal, Fermat, the Bernoulli brothers, Bayes, Laplace, Poisson, and Gauss, who are among the better known names from the 17th–19th centuries.

This emerging theory of probability became inadequate for 20th century advancement. The Russian mathematician Kolmogorov successfully developed an axiomatic model, which relied heavily on the use of set theory. The axioms provided rules to which probabilities must conform, but no guidance on assigning probabilities. Of course, the classical probability results do satisfy Kolmogorov's axioms and justify our attention to probability theory of equally likely outcomes in this chapter.

2.1 Experiments with Equally Likely Outcomes

Statistical practice relies heavily on the design and repetition of "experiments." Running an experiment can range from flipping a coin, to rolling a pair of dice, to drawing balls from a jar, to evaluating a new drug for treating hypertension. Classical probability can help us model the first three experiments, but not the fourth.

Lotteries such as Mega Millions and Powerball can generate winnings in the hundreds of millions of dollars. These games are amenable to methods in this chapter.

In 1969, the US Government held a lottery to assign military draft numbers for all birthdays by drawing 366 plastic balls from a deep glass jar. The number drawn for my birthday was 120; my younger brother's birthday received 350, a number that had a slim chance of being "called." The volunteer army was initiated, and neither of us was called. Pulling balls from an urn is a classical equally likely experiment; however, we will review the empirical evidence concerning the fairness of draft lottery later; see Problem 7 in Chapter 7.

Statistics: A Concise Mathematical Introduction for Students, Scientists, and Engineers, First Edition. David W. Scott.
© 2020 John Wiley & Sons Ltd. Published 2020 by John Wiley & Sons Ltd.

2.1.1 Simple Outcomes

A classical probability experiment results in one of n equally likely results, or outcomes, which we denote by $s_1, s_2, \ldots, s_n$. Intuitively, or by invoking symmetry, we assign equal probabilities to each simple outcome

$$P(s_i) \equiv p_i = \frac{1}{n} ; \quad \text{hence,} \quad \sum_{i=1}^{n} p_i = 1 \quad \text{or } 100\%. \tag{2.1}$$

By noting the value of n for various experiments, we would assign $p_i = \frac{1}{2}$ when flipping a coin, $p_i = \frac{1}{6}$ when rolling a single die, and $p_i = \frac{1}{366}$ for the draft lottery. In practice, all of these examples use actual physical objects; hence, we rely on the notion of perfect symmetry for correctness. However, a coin is not exactly symmetrical, nor is a wooden die of uniform density or shape. We may or may not care about small deviations from the probabilities given by these formulae. It is a statistical question to "test" the correctness of our probability model; we consider such tests in Section 8.3.5.

As an important aside, Definition (2.1) requires $n < \infty$. Letting the number of balls in the lottery approach ∞ does not seem practical. In particular, what does $p_i = 0, \forall i$ mean? (Recall $\forall$ means "for all.") However, if we instead begin with an American roulette wheel, $n = 38$, letting $n \to \infty$ does seem practical. Using a spinner and a circle labeled from 0 to 1, and everything in between uniformly spaced, we can read the result of each experiment (spinning the spinner) and record a number $x \in [0, 1)$ to as many decimal places as we are physically able to resolve. Individual probabilities are still "0" in the limit, but a spinner is far superior in practice than attempting to deal with a jar full of millions of labeled balls. We shall return to the question of how to cope with an infinite number of outcomes in Section 2.4.

Anticipating our use of set theory, we may represent a classical experiment with a Venn diagram; see Figure 2.1 for the roll of a single die. The actual location of the six simple outcomes in the Venn diagram is arbitrary. The rectangle includes all simple outcomes; hence, we refer to it as the *sure event* and denote it by Ω. In general, the set $\Omega = \{s_1, s_2, \ldots, s_n\}$.

2.1.2 Compound Events and Set Operations

The outcome of a single experiment is one and only one of the simple outcomes. In general, each outcome can be represented as a set with a single element:

$$A_1 = \{s_1\}, A_2 = \{s_2\}, \ldots, A_n = \{s_n\} .$$

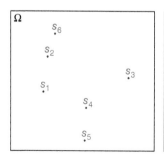 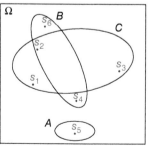

Figure 2.1 (Left) Venn diagram of the classical probability experiment rolling a single die where $n = 6$. (Right) Events A, B, and C are shown as ellipses enclosing the appropriate simple outcomes.

More generally, we may describe the outcome of an experiment in words. For example, the spots on a die are often referred to as "pips." When rolling a single die, the outcome may be five pips, or we may be interested when the number of pips is an even number, or when the number of pips is less than four. These three possible outcomes are denoted by the sets

$$A = \{s_5\} \qquad B = \{s_2, s_4, s_6\} \qquad C = \{s_1, s_2, s_3\} \,. \tag{2.2}$$

The sets in Equation (2.2) are depicted in the right frame of Figure 2.1. We refer to these sets as "events." A "simple event" has exactly one element, or one simple outcome. A "compound event" has more than one element. A compound event "occurs" if the observed element (simple outcome) is in its list. For example, if element $\{s_4\}$ is observed, then the compound event B occurs, but not events C or A.

Using ordinary set operations, we can generate new events. For example, $B = A_2 \cup A_4 \cup A_6$. The complementary event B^c records odd pips. The event A^c records pips other than a five spot. The event $B \cap C^c = BC^c = \{s_4, s_6\}$ captures even pips greater than three. *Note the special (shorthand) notation for intersection, omitting the $\cap$ symbol.*

For completeness, we include the null set (null event), $\varnothing$, which has probability 0. The sure event, Ω, is the complement of $\varnothing$, that is, $\varnothing^c = \Omega$. Clearly, $P(\Omega) = 1$. We also record here the important set operations encompassed by DeMorgan's laws:

$$\left(\bigcap_{i=1}^{m} A_i\right)^c = \bigcup_{i=1}^{m} A_i^c \quad \text{and} \quad \left(\bigcup_{i=1}^{m} A_i\right)^c = \bigcap_{i=1}^{m} A_i^c \tag{2.3}$$

and the distributive laws (using the intersection shorthand notation):

$$A \cap (B \cup C) = (A \cap B) \cup (A \cap C) = AB \cup AC \tag{2.4}$$

$$A \cup (B \cap C) = (A \cup B) \cap (A \cup C) \,. \tag{2.5}$$

We will review other set operations as they are encountered.

2.2 Probability Laws

We derive a number of probability results for the classical probability model with n equally likely outcomes. We begin with two events A and B defined on this space, and denote by n_A and n_B the number of simple outcomes in each respectively. Hence,

$$P(A) = n_A/n \qquad \text{and} \qquad P(B) = n_B/n \,.$$

There are only four generic relationships possible between A and B, which are depicted in Figure 2.2. In general, there are simple outcomes in Ω, but not in the events A and B. For convenience and to avoid degeneracies, we assume every region in Figure 2.2 has at least one simple outcome. In particular, $1 \le n_A < n$, $1 \le n_B < n$, and $n_{A \cap B} \ge 1$. It follows that $n - n_{A \cup B} \ge 1$ in all four cases. One should check that $n - n_{A \cup B} = n_{(A \cup B)^c}$.

2.2.1 Union and Intersection of Events A and B

Define the events $C = A \cup B$ and $D = A \cap B$. Let us derive the probabilities of C and D for the four cases in Figure 2.2.

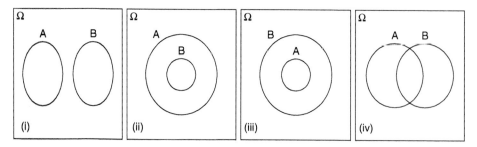

Figure 2.2 Four possible relationships between events A and B.

2.2.1.1 Case (i)

The two events A and B are disjoint, that is, $A \cap B = \varnothing$. Thus we have $n_D = 0$ and $P(D) = 0$. Also

$$n_C = n_A + n_B \implies$$

$$P(C) = \frac{n_C}{n} = \frac{n_A + n_B}{n} = \frac{n_A}{n} + \frac{n_B}{n} ;$$

hence, $\boxed{P(A \cup B) = P(A) + P(B), \quad \text{if } A \cap B = \varnothing,}$ (2.6)

or $\boxed{P(A \uplus B) = P(A) + P(B),}$ equivalently, (2.7)

using the special symbol $\uplus$ to denote the disjoint union of sets.

2.2.1.2 Cases (ii) and (iii)

We derive the outcome for case (ii), noting that case (iii) will be similar. Since $C = A \cup B = A$ and $D = A \cap B = B$, then $P(C) = P(A)$ and $P(D) = P(B)$.

2.2.1.3 Case (iv)

This is the most general case. For event D, we have

$$P(D) = \frac{n_D}{n} = \frac{n_{A \cap B}}{n} .$$

Of more interest is the result that follows from counting

$$n_C = n_A + n_B - n_{A \cap B} \implies$$

$\boxed{P(A \cup B) = P(A) + P(B) - P(A \cap B), \text{ in general.}}$ (2.8)

Intuitively, $P(A \cap B)$ is included in both $P(A)$ and $P(B)$; hence, the final term in Equation (2.8) removes the double counting. Observe that this formula also covers case (i) since $P(A \cap B) = 0$ in that case.

2.2.2 Conditional Probability

Statistics is about understanding and managing uncertainty. New information can and should influence the likelihood that an event occurs. The notation $P(B|A)$ should be read *the probability event B occurs, given that event A has occurred.* We are interested in the change from the *unconditional probability* $P(B)$ to the *conditional probability* $P(B|A)$.

2.2.2.1 Definition of Conditional Probability

Intuitively, knowing that A has occurred means that, instead of n equally likely outcomes in play, there are now only n_A possible simple outcomes. Among those n_A possibilities, there are $n_{B \cap A}$ that would result in the event B occurring. Hence, we may compute

$$P(B|A) = \frac{n_{B \cap A}}{n_A}$$

$$= \frac{n_{B \cap A}/n}{n_A/n} \implies$$

$$\boxed{P(B|A) = \frac{P(B \cap A)}{P(A)}}. \tag{2.9}$$

Naturally, we must assume $P(A) > 0$, or equivalently, $n_A > 0$.

Example: With a single die, the event $B = \{s_1, s_2, s_3\}$ that the number of pips is 1, 2, or 3 has probability $\frac{3}{6} = \frac{1}{2}$. Let event A be that the number of pips is odd, and that event $A = \{s_1, s_3, s_5\}$ has occurred. Among the $n_A = 3$ simple outcomes, $B|A = \{s_1, s_3\}$ and $n_{B \cap A} = 2$; therefore, $P(B|A) = 2/3$; hence, $B|A$ is more likely to occur than B.

2.2.2.2 Conditional Probability With More Than Two Events

Equation (2.9) can be rewritten in an intuitive form:

$$P(A \cap B) = P(A)\, P(B|A) ;$$

that is, the fraction of time that both A and B occur simultaneously equals the fraction of time that A occurs alone times the fraction of time that B occurs *given* that A has occurred. Note we have used the "frequentist" language that probabilities are long-term occurrence fractions.

We may extend this to more than two events, for example,

$$P(A \cap B \cap C) = P(A)\, P(B|A)\, P(C|A \cap B) \tag{2.10}$$

$$P(ABCD) = P(A)\, P(B|A)\, P(C|AB)\, P(D|ABC) , \tag{2.11}$$

using our shorthand notation for intersection in formula (2.11).

Example: The birthday problem. Suppose we ask for the probability that a group of n students (labeled $1, 2, \ldots, n$) has no duplicate birthdays. Let the event A_k be that the birthday of the kth student is different from those ahead of her. Then the desired probability is

$$P(\text{no matches}) = P(A_1 \cap A_2 \cap \cdots \cap A_n) = P(A_1 A_2 \cdots A_n)$$

$$= P(A_1)P(A_2|A_1)P(A_3|A_1 A_2) \cdots P(A_n|A_1 A_2 \cdots A_{n-1})$$

$$= 1 \cdot \frac{364}{365} \cdot \frac{363}{365} \cdots \frac{365 - n + 1}{365} ,$$

assuming all birthdays in a non-leap year are equally likely; see Figure 2.3. We see only 23 students are required to have a 50–50 chance of duplicate birthdays. Given the many possibilities of duplicate birthdays, we have chosen to find the complementary probability, which has only one way of occurring. This is a common strategy as our problems become more involved.

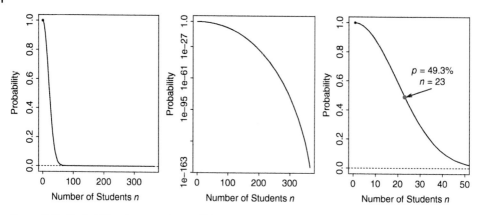

Figure 2.3 Probability that n students all have different birthdays, plotted using three different scales.

2.2.3 Independent Events

If the knowledge that A has occurred does *not* change the probability of event B, we say that events A and B are *independent*. Using Equation (2.9), we obtain the more useful characterization:

$$P(B|A) = P(B) = \frac{P(B \cap A)}{P(A)} \implies$$

$$\boxed{P(B \cap A) = P(A)P(B)} \quad \text{if } A \text{ and } B \text{ are independent.} \tag{2.12}$$

This product relationship is often taken as the definition of independent events.

Examples: In Figure 2.2, we assume both $P(A)$ and $P(B)$ are between 0 and 1. In case (i), $P(B|A) = 0$, while in case (iii), $P(B|A) = 1$; therefore, events A and B are *not* independent in either case. The same holds for case (ii), since we can prove $P(B|A) > P(B)$, noting $n_{A \cap B} = n_B$:

$$P(B|A) = \frac{n_{A \cap B}}{n_A} = \frac{n_B}{n_A} = \frac{n_B}{n} \times \frac{n}{n_A}$$

$$= P(B) \times \frac{n}{n_A} > P(B) . \tag{2.13}$$

This assumes that $0 < n_A < n$ of course. Thus with the exception of some degenerate cases, independence can occur only in case (iv).

With more than two events, independence is defined as

$$P(ABC) = P(A)\, P(B)\, P(C)$$

$$P(ABCD) = P(A)\, P(B)\, P(C)\, P(D) , \quad \text{or in general}$$

$$\boxed{P\left(\bigcap_{i=1}^{n} A_i\right) = \prod_{i=1}^{n} P(A_i)} \iff \{A_1, \dots, A_n\} \text{ are independent.} \tag{2.14}$$

2.2.4 Bayes Theorem

Is there a relationship between the two conditional probabilities $P(B|A)$ and $P(A|B)$? The answer is "yes." The explicit formula is called Bayes theorem, a result famously published posthumously. An algebraic proof simply applies a rearranged version of Equation (2.9) twice:

$$P(A)P(B|A) = P(B \cap A) = P(A \cap B) = P(B)P(A|B) \; ;$$

hence, we obtain

$$\boxed{P(A|B) = \frac{P(B|A)P(A)}{P(B)} \quad \text{Bayes theorem.}} \tag{2.15}$$

Applying (2.15) is often difficult because $P(B)$ is unknown or baffling to compute. We will derive a useful solution soon.

Corollary: If A does not change the probability of B, is the reverse true as well? That is, if B is independent of A, i.e. $P(B|A) = P(B)$, is A necessarily independent of B? The answer is yes. Using Bayes theorem,

$$P(A|B) = \frac{P(B|A)P(A)}{P(B)}$$

$$= \frac{P(B)P(A)}{P(B)}$$

$$= P(A) \quad \square.$$

Example: Bayes theorem is commonly used for medical diagnosis. Suppose C is the event that Jack has prostate cancer, and T is the event that a protein-specific antigen (PSA) test is "positive" for cancer. If Jack has a PSA test and the result is "positive," what is the probability that Jack has prostate cancer?

Thus we seek $P(C|T)$ for men like Jack. According to the 2006 Prostate Cancer Prevention Trial (Thompson et al. (2006)) among men aged 55 or older,

$$P(T|C) = 0.205$$

$$P(T^c|C^c) = 0.938 \; ;$$

thus the PSA test has surprisingly low sensitivity (missing almost 80% of cancers), but adequate specificity (giving a false positive result for 6.2% of cancer-free men). In addition, the Centers for Disease Control states that a man 60–70 years old has a 0.058 probability of having prostate cancer.

Thus, we can use this epidemiologically derived evidence and Bayes theorem to find the desired conditional probability:

$$P(C|T) = \frac{P(T|C)P(C)}{P(T)}$$

$$= \frac{0.205 \times 0.058}{P(T)} \; .$$

Since $P(T)$ is not given explicitly, we note that there are two ways to obtain a positive test, either having or not having cancer, i.e. $T = TC \uplus TC^c$. Thus, $P(T) = P(TC) + P(TC^c) = P(T|C)P(C) + P(T|C^c)P(C^c)$. We computed the first term in the numerator; the second is given by $(1 - .938) \times (1 - .058)$, noting $P(T|C^c) = 1 - P(T^c|C^c)$ (see Problem 3), giving

$$P(C|T) = 16.91\% \quad \text{for men like Jack.} \tag{2.16}$$

This surprising result is a fact of life among much of medical screening and diagnosis. Note that the PSA result Equation (2.16) almost tripled the probability Jack has cancer; however, the probability is not high enough to be directly actionable. Further testing or watchful waiting might be warranted.

Remark: The situation is surprisingly similar for women who undergo breast exams and mammograms. A cost–benefit analysis approach is logical but frustratingly imprecise to advocates for yearly screenings.

Note if 1,000,000 men aged 60–70 were given a PSA test, there would be 11,890 positive PSA tests among the 58,000 men with prostate cancer, but there would also be many more positive tests (58,404) among the 942,000 men without prostate cancer. There are two general strategies to improve the situation: (1) limit the test to "high risk" individuals (using family history, for example), so that $P(C)$ is much greater than 0.058; or (2) replace the PSA with a test with greater sensitivity and specificity (but how much more expensive)? We examine these two options in Problem 4.

2.2.5 Partitions and Total Probability

A collection of m events $A_1, A_2, \ldots, A_m$ is a partition of a space Ω, if they are pairwise disjoint and their union is Ω; that is,

$$\text{Partition of } \Omega: \quad A_i \cap A_j = \varnothing, i \neq j, \quad \text{and} \quad \bigcup_{i=1}^{m} A_i = \biguplus_{i=1}^{m} A_i = \Omega.$$

We assume $n_{A_i} > 0$ to avoid degeneracies, such as division by 0. By construction, $\sum_{i=1}^{m} n_{A_i} = n$. See the left frame of Figure 2.4 for an example.

Next, we demonstrate how the probability of an event, B, may be computed piecewise, using the partition $\{A_i\}$. The middle frame of Figure 2.4 shows an example of the interaction

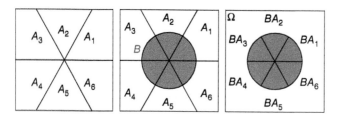

Figure 2.4 (Left) Venn diagram of partition of Ω into $m = 6$ sets $A_1, A_2, \ldots, A_6$; (middle) set B superimposed upon partition; (right) set B decomposed into m disjoint events using the partition. In this Figure we use the shorthand notation for intersection, namely, $BA_j = B \cap A_j$.

of the set B and the partition. Algebraically for a general partition of m sets, we have

$$B = B \cap \Omega$$

$$= B \cap \left(\bigcup_{i=1}^{m} A_i \right)$$

$$= \bigcup_{i=1}^{m} (B \cap A_i) \qquad \text{by the distributive law.} \tag{2.17}$$

Thus the probability of B may be computed by summing its portion in each partition set, and then using Bayes theorem for each:

$$n_B = \sum_{i=1}^{m} n_{BA_i}$$

$$\frac{n_B}{n} = \sum_{i=1}^{m} \frac{n_{BA_i}}{n}$$

$$P(B) = \sum_{i=1}^{m} P(B \cap A_i)$$

$$\boxed{P(B) = \sum_{i=1}^{m} P(B|A_i)P(A_i) \qquad \text{law of total probability.}} \tag{2.18}$$

The *law of total probability* allows us to divide the problem of computing $P(B)$ into m smaller (and presumably easier) subproblems, as we did with our Bayes theorem prostate cancer diagnosis example.

2.3 Counting Methods

Classical probability relies on tabulating simple outcomes in order to compute probabilities. In this section, we derive some of the more important counting formulae. We generally assume that we select all or a subset of n distinguishable (or ordered) objects.

Fundamental principal of counting (FPC): If an experiment can be broken down into a finite sequence of m selection processes

$$S = S_1 \times S_2 \times \cdots \times S_m$$

and if the number of selections at the kth step does *not* depend on the preceding selections, then

$$n(S) = n(S_1) \times n(S_2) \times \cdots \times n(S_m) = \prod_{\ell=1}^{m} n(S_\ell) . \tag{2.19}$$

Example: Throw a die twice: $n(S) = 6 \times 6 = 36$.

Example: Make selections in a three-step process; see Figure 2.5.

Hint: Note the use of the phrase "for each" in this example. That is the clue to use the multiplication rule; see Figure 2.5.

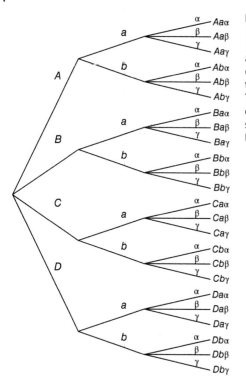

Figure 2.5 Illustration of the FPC. Select one of $\{A, B, C, D\}$, then one of $\{a, b\}$, and finally one of $\{\alpha, \beta, \gamma\}$. For each selection at the first step, there are two choices at the second step. Finally, for each of the selections after the first two steps, there are three choices at the third step. Therefore, $n(S) = 4 \times 2 \times 3 = 24$ possibilities. It is common to use a tree diagram to visualize the selection process, recording the selections on the leaves.

In the following sections, we assume we are selecting a subset of size r from n distinguishable objects.

2.3.1 With Replacement

At the kth step, there are still n objects available. Therefore, after r steps,

$$n(S) = n^r .$$

2.3.2 Without Replacement (Permutations)

There are n objects available at the first step. For each choice made at the first step, there are only $n - 1$ objects available at the second step, and so on. After the rth step,

$$n(S) = n \times (n - 1) \times (n - 2) \times \cdots \times (n - r + 1) = \frac{n!}{(n - r)!} .$$

This defines the number of **permutations** of r objects selected from n objects without replacement, which we will denote by nP_r. Then

$$\boxed{{}^nP_r = \frac{n!}{(n - r)!}} ; \quad \text{in particular,} \quad \boxed{{}^nP_n = n!} . \tag{2.20}$$

Recall that 0! is defined to be equal to 1.

2.3.3   Without Replacement or Order (Combinations)

In this case, we consider the six permutations $(1, 2, 3), (1, 3, 2), (2, 1, 3), (2, 3, 1), (3, 1, 2)$, and $(3, 2, 1)$ to be identical. In general, each permutation appears $r!$ times. We denote by ^nC_r or $\binom{n}{r}$ the number of selections of r objects from n objects, where we do not distinguish on the order of the objects. Hence,

$$^nC_r = \binom{n}{r} = \frac{^nP_r}{r!} \text{ ; specifically } \boxed{^nC_r = \binom{n}{r} = \frac{n!}{(n-r)!\, r!}}. \tag{2.21}$$

2.3.4   Examples

Cabinet members: Select 4 of 12 cabinet members as officers. Here we need to distinguish the office for which a candidate is selected. This is accomplished by recording the order the 4 were selected:

$$^{12}P_4 = \frac{12!}{8!} = 12 \cdot 11 \cdot 10 \cdot 9 = 11{,}880.$$

Note there are 12 choices for the president, 11 for the vice-president, then 10 for the treasurer, and finally 9 for the secretary, for example.

Poker hands: Select five cards from a full and well-shuffled deck:

$$^{52}C_5 = \binom{52}{5} = \frac{52 \cdot 51 \cdot 50 \cdot 49 \cdot 48}{5!} = 2{,}598{,}960.$$

This is the number of unique poker hands in five-card draw. Note that while we may arrange the five cards we draw, this does not affect the number of poker hands. All of these are equally likely. Alternatively, it is not wrong to consider the permutations as equally likely, just generally a waste of time (unless we are peeking at the cards as we are dealt them!). If we retain order information in the denominator, we must count appropriately in the numerator when computing probabilities.

Some specific poker hands, with $n = \binom{52}{5}$ equally likely outcomes:

$$n(\text{royal flush}) = \underbrace{\binom{4}{1}}_{\text{select suit}} \times \underbrace{\binom{5}{5}}_{\text{select AKQJ10}} = 4$$

$$n(\text{4 aces}) = \underbrace{\binom{4}{4}}_{\substack{\text{all aces}}} \times \underbrace{\binom{4}{1}}_{\substack{\text{suit for} \\ \text{last card}}} \times \underbrace{\binom{12}{1}}_{\substack{\text{number of} \\ \text{last card}}} = 48$$

$$n(\text{2 pair}) = \underbrace{\binom{13}{2}}_{\substack{\text{numbers} \\ \text{for pairs}}} \times \underbrace{\binom{4}{2}\binom{4}{2}}_{\text{suits for pairs}} \times \underbrace{\binom{11}{1}\binom{4}{1}}_{\substack{\text{number and} \\ \text{suit for last card}}} = 123{,}552.$$

Note that if we started counting two pair with $\binom{13}{3}$ and got the unordered list $\{A, 10, 3\}$, we would not know which numbers were to get the pairs and which number was not. Thus, for each triple, we would need to select the numbers to get the pairs, or which number would not. These would be $\binom{3}{2}$ and $\binom{3}{1}$, respectively. Both equal 3. The probability of getting two pair is 4.75%.

Finally, suppose we had counted two pair with the formula $^{13}P_2 \times \binom{4}{2}\binom{4}{2} \times 44$. Then among the leaves of the tree, we would end up with both

$$\{A_\spadesuit, A_\heartsuit, 4_\diamondsuit, 4_\heartsuit, 8_\spadesuit\} \quad \text{and} \quad \{4_\diamondsuit, 4_\heartsuit, A_\spadesuit, A_\heartsuit, 8_\spadesuit\}$$

appearing as different hands. But these are identical if we take the n equally likely hands to be $\binom{52}{5}$. Permutations are useful when we wish to distinguish objects by the device of ordering them. Here, that is inappropriate.

2.3.5 Extended Combinations (Multinomial)

Assign everyone to one and only one of k groups; thus $\sum_{i=1}^{k} n_i = n$. Then $n(S)$ is clearly

$$n(S) = \binom{n}{n_1}\binom{n-n_1}{n_2}\binom{n-n_1-n_2}{n_3}\cdots\binom{n-n_1-n_2-\cdots-n_{k-1}}{n_k}.$$

Expanding this formula, there is a lot of cancellation. The resulting **multinomial combination coefficient** is given by

$$\binom{n}{n_1,\,n_2,\,\ldots,\,n_k} = \frac{n!}{n_1!n_2!\cdots n_k!}. \tag{2.22}$$

Example: Twelve freshmen are to be assigned to two four-person rooms and two two-person rooms. The number of possible groupings is

$$\binom{12}{4,\ 4,\ 2,\ 2} = \frac{12!}{4!\,4!\,2!\,2!} = 207,900.$$

2.4 Countable Sets: Implications as $n \to \infty$

Extending a finite set of outcomes to an infinite set is not always straightforward. We give two examples and introduce the concept of countability.

2.4.1 Selecting Even or Odd Integers

Consider the following question: what is the probability a random positive integer is odd? The intuitive answer is 50%, which follows from extending the following sequence where n is even, and odd integers highlighted in bold:

1 2 **3** 4 **5** 6 **7** 8 **9** 10 **11** 12 $\cdots$ **n − 1** n

and letting $n \to \infty$. However, consider the following sequences:

1 2 4 **3** 6 8 **5** 10 12 **7** 14 16 **9** 18 20 **11** 22 24 **13** $\cdots$

1 2 4 6 **3** 8 10 12 **5** 14 16 18 **7** 20 22 24 **9** 26 28 30 **11** $\cdots$

1 2 4 6 8 **3** 10 12 14 16 **5** 18 20 22 24 **7** 26 28 30 32 **9** $\cdots$.

For each, as the length of the sequence goes to ∞, all positive integers are included, but the probabilities of an odd integer are approximately 1/3, 1/4, and 1/5, respectively. Apparently, we can create a sequence that has "limiting" probability any rational number p/q, where p and q are integers and $p \le q$. Thus selecting a random positive integer is not a well-defined experiment.

2.4.2 Selecting Rational Versus Irrational Numbers

Consider our spinner example from Section 2.1.1 that results in a random number, x, on the interval $[0, 1]$. Here the question is: What is the probability x is a rational number (versus the converse that x is irrational)? Clearly, there are an infinite number of rational numbers and an infinite number of irrational numbers; however, since every number between 0 and 1 is (supposedly) equally likely, the answer might be 50%.

Consider the following sequence of rational numbers:

$$\frac{0}{1} \; \frac{1}{1} \quad \frac{0}{2} \; \frac{1}{2} \; \frac{2}{2} \quad \frac{0}{3} \; \frac{1}{3} \; \frac{2}{3} \; \frac{3}{3} \quad \frac{0}{4} \; \frac{1}{4} \; \frac{2}{4} \; \frac{3}{4} \; \frac{4}{4} \quad \frac{0}{5} \; \frac{1}{5} \; \frac{2}{5} \; \frac{3}{5} \; \frac{4}{5} \; \frac{5}{5} \quad \cdots \; .$$

Every rational number between 0 and 1 is included (at least once) in this list. (We can delete duplicates, of course.) Thus the set of rational numbers on $[0, 1]$ satisfies an important property.

> **Countably infinite sets:** A countable set is either finite or countably infinite, in which case there is a one-to-one association of elements of the set and the positive integers $\{1, 2, 3, \dots\}$.

In the 19th century, Georg Cantor proved that the real numbers are uncountable. We may conclude that the irrational numbers between 0 and 1 are uncountably infinite. In an advanced real analysis course, we would say the "measure" of rational numbers on $[0, 1]$ is 0, and the "measure" of irrational numbers is 1. Hence, the probability of selecting a rational number is 0.

Many of the technical problems we shall encounter with infinity can be addressed if we limit ourselves to countably infinite sets. In other situations, we can use the continuity of the real line and the fact that any irrational number can be approximated by a sequence of rational numbers. However, in general, we leave these technical details to subsequent courses dealing with measure theory.

Finally, we note that it is straightforward to extend our construction to rational numbers of $[0, 2]$, then all positive rational numbers, and by alternating signs, to negative rational numbers. Thus the rational numbers on the real line are countable.

2.5 Kolmogorov's Axioms

Kolmogorov laid out a minimal set of rules that a function P assigning a "probability" to an event E must satisfy. We assume the existence of a sample space Ω. An event E is a subset of Ω. Let $\mathscr{F}$ denote the list of all events. (Technically, $\mathscr{F}$ is an algebra of sets, i.e. (1) $\Omega \in \mathscr{F}$; (2) $E^c \in \mathscr{F}$ if $E \in \mathscr{F}$; and (3) $E_1 \cap E_2 \in \mathscr{F}$ if $E_1 \in \mathscr{F}$ and $E_2 \in \mathscr{F}$.)

Kolmogorov's axioms of probability: Suppose we are given a sample space Ω, a list of all events $\mathcal{F}$, and a function P that assigns a probability $P(E)$ to every set $E \in \mathcal{F}$. Then the probability function P must satisfy:

Axiom I $P(E) \in \mathbb{R}^1$ and $P(E) \geq 0, \forall E \in \mathcal{F}$

Axiom II $P(\Omega) = 1$

Axiom III For a finite set of disjoint events, $\{E_1, E_2, \ldots, E_n\}$,

$$P\left(\biguplus_{i=1}^{n} E_i\right) = \sum_{i=1}^{n} P(E_i) \tag{2.23}$$

Axiom III' For a countably infinite set of disjoint events, $\{E_1, E_2, \ldots\}$,

$$P\left(\biguplus_{i=1}^{\infty} E_i\right) = \sum_{i=1}^{\infty} P(E_i) . \tag{2.24}$$

It follows that

$$P(\varnothing) = 0,$$

$$\text{if } A \subseteq B, \text{ then } P(A) \leq P(B) , \quad \text{and}$$

$$0 \leq P(E) \leq 1 .$$

To prove the second assertion, note that if $A \subseteq B$, then

$$B = (AB) \uplus (A^c B) = A \uplus (A^c B) \text{ ; hence,}$$

$$P(B) = P(A) + P(A^c B) . \quad \text{Therefore,}$$

$$P(B) \geq P(A) , \text{ since } P(A^c B) \geq 0 .$$

The other two assertions follow immediately, as do all of the probability results we derived in Sections 2.2 in the equally likely scenario. Of course, the general results must include the classical probability formulae.

2.6 Reliability: Series Versus Parallel Networks

We analyze two simple systems comprised of n independent components, $\{A_1, A_2, \ldots, A_n\}$, that operate with probabilities $p_i = P(A_i), i = 1, \ldots, n$. The first system places the components in series, while the second places the components in parallel; see Figure 2.6. We wish to compute the probability that there is a working path connecting the input and output. Let S denote the event that a working path exists.

2.6.1 Series Network

In a series configuration, all n components must operate for the system to work, call this event S. In set notation, the event the S is defined by

$$S = A_1 \cap A_2 \cap \cdots \cap A_{n-1} \cap A_n .$$

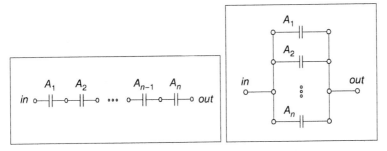

Figure 2.6 A series network (left) and parallel network (right) of n components.

Since the components are independent

$$P(S) = P(A_1 \cap A_2 \cap \cdots \cap A_{n-1} \cap A_n), \quad \text{or}$$

$$\boxed{P(S) = \prod_{i=1}^{n} p_i, \qquad \text{series network.}}$$

Example: A new homeowner lives in a 100-year flood plain. If she took a 30-year mortgage, what is the probability she will not experience a 100-year flood over the life of the mortgage? **Answer:** If we assume each year is independent of the others, then the probability of no flood is $p_i = 1 - 0.01$ for each year; hence, $P(S) = 0.99^{30} = 74.0\%$. Better buy flood insurance.

Remark: Since all n components must work for the system to function, a series network cannot be more reliable than its weakest component, that is, $P(S) \le \min_i(p_i)$. If $p_i = p$ for all n components, then to achieve a reliability level of 99%, p must satisfy $p^n \ge 0.99$ or $p \ge 0.99^{1/n}$. For 30 years, $p > 0.99966$ must be satisfied.

2.6.2 Parallel Network

On the other hand, in a parallel configuration, only one (or more) of the n components must operate for the system to work. The system will fail only if all n components fail simultaneously. In set notation, the system works is

$$S = A_1 \cup A_2 \cup \cdots \cup A_{n-1} \cup A_n, \quad \text{or}$$

$$S^c = \left(\bigcup_{i=1}^{n} A_i \right)^c = \bigcap_{i=1}^{n} A_i^c.$$

Since the events $\{A_i\}$ are independent, so are the complementary events, $\{A_i^c\}$; hence, $P(\cap A_i^c) = \prod P(A_i^c)$. Continuing,

$$P(S) = 1 - P(S^c) = 1 - \prod_{i=1}^{n} P(A_i^c), \quad \text{or}$$

$$\boxed{P(S) = 1 - \prod_{i=1}^{n} (1 - p_i), \qquad \text{parallel network.}}$$

Remark: Since only one of the n components need work for the system to function, a parallel network will be even more reliable than its strongest component; that is, $P(S) \geq \max_i(p_i)$. If $p_i = p$ for all n components, then to achieve a reliability level of 99%, p must satisfy

$$1 - (1 - p)^n \geq 0.99$$
$$(1 - p)^n \leq 1 - 0.99$$
$$1 - p \leq 0.01^{1/n} \quad \text{or}$$
$$p \geq 1 - 0.01^{1/n} \, .$$

For example, when $n = 5$, $p \geq 0.602$. For ten components, p need only be 0.369. In practice, an engineer seeking redundancy might design a few expensive and highly reliable components, together with several less reliable but cheaper backups. The worry is that in a catastrophic failure, the independence assumption may not hold.

Problems

2.1 Show that the formula (2.8) for case (iv) also covers cases (i)–(iii).

2.2 How does $P(T|C)$ change depending on increasing $P(C)$ versus the PSA test characteristics?

2.3 Show $P(A|B) + P(A^c|B) = 1$. Similarly, show $P(A|B^c) + P(A^c|B^c) = 1$.

2.4 Examine how $P(C|T)$ improves as a function of $P(C)$, $P(T|C)$, and $P(T^c|C^c)$.

2.5 Prove these and/or draw Venn Diagrams: Equations (2.4) and (2.5).

2.6 Show that for case (ii), $P(B|A)$ in Equation (2.13) can be expressed in terms of $P(A)$ and $P(B)$ to reach the same conclusion.

2.7 Show formula (2.22) is correct.

2.8 Prove that a series network is no more reliable than its weakest component, while a parallel network is more reliable than its strongest component.

3

Random Variables and Models Derived From Classical Probability and Postulates

The examples in the previous chapter may be described as coming from a **discrete uniform distribution**, which is defined as an experiment with n equally likely outcomes (n finite). In this chapter, we begin to explore other distributions. There are, in fact, an infinite number of possible probability distributions. We will focus on some of the most important and common examples. These distributions all share the feature that they follow from simple rules and intuitive motivation. In addition, we explore how the results of experiments are measured and recorded.

3.1 Random Variables and Probability Distributions: Discrete Uniform Example

While a verbal description of an event and its occurrence (or non-occurrence) is a good starting point, it will prove much more powerful to use a numerical value to record the result of an experiment. For our example in Figure 2.1, if the simple outcome is s_2, the events B and C occurred, but event A failed to occur. More generally, the result of an experiment might be either yes/no or true/false or success/failure. We might assign the integer values 1 and 0, respectively, to those results.

We might also use the capital letter, X, to represent the possible results. In general, X is a function of the event, s, and records a numerical value as $X(s)$, which may be any real number. X is called a **random variable** (r.v.), a name that can engender some confusion, for X precisely records the result of a random experiment; that is, the randomness is not in the measurement but in the experiment being run. Then after the experiment is run, a lower case x is used to denote and record a specific result (or "outcome"). For our example in Figure 2.1, if the simple outcome is $\{s_2\}$, the events A, B, or C are recorded as $x = 0, x = 1$, or $x = 1$, respectively.

To summarize our new notation, $X(s) = x$. Following convention, we often suppress the functional notation of the random variable, $X(\cdot)$, and just write X alone. We illustrate our ideas for the discrete case in the next section, and for the continuous case in Section 3.2.

Statistics: A Concise Mathematical Introduction for Students, Scientists, and Engineers, First Edition. David W. Scott.
© 2020 John Wiley & Sons Ltd. Published 2020 by John Wiley & Sons Ltd.

3.1.1 Toss of a Single Die

To make these ideas more concrete, we consider two experiments: one rolling a single die and a second rolling a pair of dice. In the first experiment, consider the random variable, X, that records the number of the pips observed on the toss of a single die. X takes as its value one and only one of the integers $1, 2, \ldots, 6$. These values are equally likely, that is, the probability of observing the value, i, is $p_i = P(X = i) = 1/6$, $i = 1, 2, \ldots, 6$.

There are two important functions that statisticians use to describe the distribution of such probabilities:

$$F(x) = P(X \leq x) \quad \textbf{cumulative distribution function} \quad \text{(CDF)} \tag{3.1}$$

$$p(x) = P(X = x) \quad \textbf{probability mass function} \quad \text{(PMF).} \tag{3.2}$$

These are easily computed for our example: see Table 3.1. The CDF and PMF are plotted in Figure 3.1.

3.1.2 Toss of a Pair of Dice

The second experiment, rolling a pair of dice, is slightly more complicated. Here, the random variable, X, records the sum of the pips (numbers) on two dice. Clearly, X can take as its value one and only one of the integers $2, 3, \ldots, 12$. But these values are not equally

Table 3.1 Probabilities for roll of a single die.

x	1	2	3	4	5	6
$F(x)$	$\frac{1}{6}$	$\frac{2}{6}$	$\frac{3}{6}$	$\frac{4}{6}$	$\frac{5}{6}$	$\frac{6}{6}$
$p(x)$	$\frac{1}{6}$	$\frac{1}{6}$	$\frac{1}{6}$	$\frac{1}{6}$	$\frac{1}{6}$	$\frac{1}{6}$

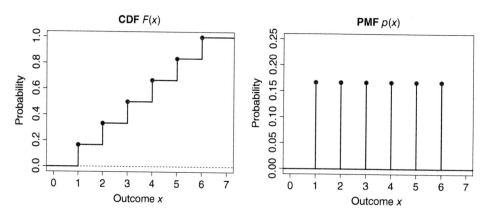

Figure 3.1 (Left) The cumulative distribution function for the roll of a single die; and (right) its probability mass function.

Table 3.2 The 36 equally likely simple outcomes when rolling a pair of dice.

$(1,1)$	$(1,2)$	$(1,3)$	$(1,4)$	$(1,5)$	$(1,6)$
$(2,1)$	$(2,2)$	$(2,3)$	$(2,4)$	$(2,5)$	$(2,6)$
$(3,1)$	$(3,2)$	$(3,3)$	$(3,4)$	$(3,5)$	$(3,6)$
$(4,1)$	$(4,2)$	$(4,3)$	$(4,4)$	$(4,5)$	$(4,6)$
$(5,1)$	$(5,2)$	$(5,3)$	$(5,4)$	$(5,5)$	$(5,6)$
$(6,1)$	$(6,2)$	$(6,3)$	$(6,4)$	$(6,5)$	$(6,6)$

(x_1, x_2):

likely. Since *for each* pip on the first die, there are six equally likely possible outcomes on the second die, there must be 6×6 equally likely outcomes; hence, $n = 36$ is the correct denominator.

We may record these outcomes as an ordered pair, (x_1, x_2), where the entries refer to the first and second die, respectively; see Table 3.2. (If instead we imagined viewing the entries as unordered, then there would be 21 unique simple outcomes. For example, entries $(1, 2)$ and $(2, 1)$ in Table 3.2 would be considered the same; however, these 21 outcomes again would not be equally likely.) In fact, we can introduce two random variables X_1 and X_2, where X_1 records the number of pips on the first die, X_2 on the second die; then $X = X_1 + X_2$ models the total. Note X, X_1, and X_2 are three distinct r.v.s.

Since $n = 36$, it is straightforward to compute the probabilities for the random variable, X, by counting the relevant ordered pairs in Table 3.2 and dividing by n; see Table 3.3 and Figure 3.2. Again, note that $\sum_{x=2}^{12} p(x) = 1$.

In summary, we have introduced random variables as a mechanism to record the outcome of an experiment in a numerical format. Again, the upper-case and lower-case conventions, X and x, for random variables can best be understood by thinking of the random variable as a function of the simple outcome, s, namely

$$X(s) = x \;;$$

that is, the random variable function X maps the simple outcome s to a particular value $x \in \mathbb{R}$.

Remark: In an advanced mathematical probability and statistics course, we will learn that a random variable must be a "measurable function." We will assume that our work at this level does not require this concept.

Table 3.3 Probability distribution function for the sum of pips on two dice.

x	2	3	4	5	6	7	8	9	10	11	12
$F(x)$	$\frac{1}{36}$	$\frac{3}{36}$	$\frac{6}{36}$	$\frac{10}{36}$	$\frac{15}{36}$	$\frac{21}{36}$	$\frac{26}{36}$	$\frac{30}{36}$	$\frac{33}{36}$	$\frac{35}{36}$	$\frac{36}{36}$
$p(x)$	$\frac{1}{36}$	$\frac{2}{36}$	$\frac{3}{36}$	$\frac{4}{36}$	$\frac{5}{36}$	$\frac{6}{36}$	$\frac{5}{36}$	$\frac{4}{36}$	$\frac{3}{36}$	$\frac{2}{36}$	$\frac{1}{36}$

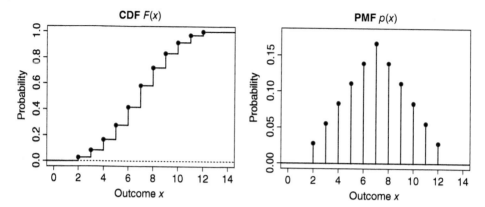

Figure 3.2   (Left) The cumulative distribution function for the sum of pips on two dice; and (right) its probability mass function. From the shape of the PMF, this is a **discrete isosceles triangular distribution**.

3.2   The Univariate Probability Density Function: Continuous Uniform Example

Next we consider the random variable recording the result of our spinner example introduced in Section 2.1.1. We can use intuitive idea of **geometric probability** to compute the cumulative distribution function. For example, the probability the spinner lands in the interval $(0.2, 0.6)$ is 40%, which is the length of the interval relative to the whole interval $[0, 1]$. Thus, if the interval is $[0, x]$, then the probability is $100 \cdot x\%$ for $0 \le x \le 1$. Thus, the CDF is given by

$$F(x) = P(X \le x) = \begin{cases} 0 & \text{if } x < 0 \\ x & \text{if } 0 \le x \le 1 \\ 1 & \text{if } x > 1 \,. \end{cases} \tag{3.3}$$

Aside: Since the probability the spinner lands on any particular value of x, say 0, $\frac{1}{4}$, or 1, is **zero**, we may treat the intervals $(0, 1)$, $[0, 1)$, $(0, 1]$, and $[0, 1]$ as indistinguishable as a practical matter.

Definition: If all of the values on an interval (a, b) are equally likely, then we say the probability law follows a **uniform distribution**, which we denote as $\text{Unif}(a, b)$.

Furthermore, if the CDF, $F(x)$, is a differentiable function so that $F'(x)$ exists, we may use the fundamental theorem of calculus to write

$$P(X \le x) = F(x) = \int_{-\infty}^{x} \frac{dF(t)}{dt} dt \,, \tag{3.4}$$

since $F(-\infty) = 0$. This allows us to extend the definition of the PMF given in Equation (3.2) for a discrete uniform distribution to a continuous distribution. We define the **probability**

density function (PDF) to be the derivative of the CDF:

$$f(x) = \frac{d}{dx} F(x) \qquad \boxed{\text{PDF for a continuous r.v.}} \tag{3.5}$$

If we know the PDF, we can compute the CDF via

$$F(x) = P(X \le x) = P(X \in (-\infty, x]) = \int_{-\infty}^{x} f(t)\, dt\,, \tag{3.6}$$

again by the fundamental theorem of calculus. Note the use of the dummy argument t to avoid confusion over multiple uses of the symbol x.

For the spinner CDF in Equation (3.3), the probability density function is

$$f(x) = \begin{cases} 0 & \text{if } x < 0 \\ ? & \text{if } x = 0 \\ 1 & \text{if } 0 < x < 1 \\ ? & \text{if } x = 1 \\ 0 & \text{if } x > 1\,. \end{cases} \tag{3.7}$$

The derivative of the CDF at $x = 0$ and $x = 1$ is not well defined; hence, we can assign any convenient values. If we consider the spinner interval to be $[0, 1]$, we can choose the value 1; see Figure 3.3.

Notes:

1. For the spinner example, the outcomes 0 and 1 are the same, so $[0, 1]$ is not a good choice. Use $[0, 1)$ or $(0, 1]$ to avoid the collision.
2. Using the same function, $f(x)$, to describe both the PMF and PDF in the discrete and continuous cases might not cause much confusion. In fact, Figures 3.1 and 3.3 share a lot of features. (For engineers, the idea that the derivative of the staircase CDF in the left frame Figure 3.1 exists as a Dirac delta function shown in the right frame would support this notational choice.) However, we will use $p(x)$ and $f(x)$ to distinguish the discrete and continuous cases, respectively.

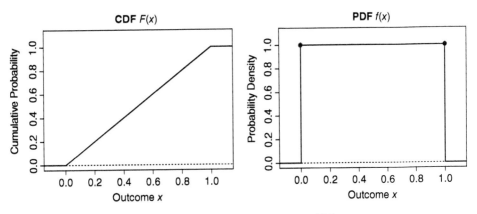

Figure 3.3   (Left) The CDF for a Unif(0, 1) density; and (right) its PDF.

3. The CDF $F(x)$ is a non-decreasing function of x, with $F(-\infty) = 0$ and $F(\infty) = 1$. In the continuous case, the PDF must be non-negative since $f = F'$. A probability law need satisfy only two conditions:

$$p(x) \geq 0 \quad \text{and} \quad \sum_x p(x) = 1 \text{ (discrete)} \tag{3.8}$$

$$f(x) \geq 0 \quad \text{and} \quad \int_{-\infty}^{\infty} f(x)\, dx = 1 \text{ (continuous)}. \tag{3.9}$$

3.2.1 Using the PDF to Compute Probabilities

In fact, the PDF can be used to compute the probability of any event, A, in particular, for intervals $A = (a, b]$: since $(-\infty, a]$ and $(a, b]$ are disjoint,

$$P(X \in (a, b]) = P(X \leq b) - P(X \leq a)$$

$$= F(b) - F(a)$$

$$= \int_{x=-\infty}^{b} f(x)\, dx - \int_{x=-\infty}^{a} f(x)\, dx \quad \text{or}$$

$$P(X \in A) = \int_{x=a}^{b} f(x)\, dx = \int_A f(x)\, dx. \tag{3.10}$$

Since $P(X = a) = 0$ for a continuous r.v., Equation (3.10) gives the probability for all four of the intervals (a, b), $[a, b)$, $(a, b]$, and $[a, b]$.

In general, the set A can be any generalized interval, which is generated by a *countable union, intersection, or complement* of open intervals (a, b). Note, for example, that

$$[a, b] = \bigcap_{n=1}^{\infty} \left(a - \frac{1}{n}, b + \frac{1}{n} \right),$$

so that closed intervals are included as well. In an advanced class, these are called the **Borel algebra of sets**. Some examples using (3.10) are shown in Figure 3.4. Probabilities are "areas under the curve."

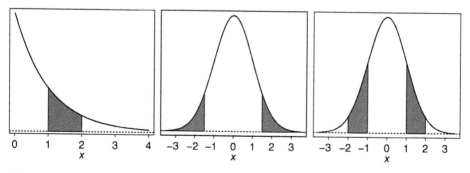

Figure 3.4 The shaded areas give the probabilities of the events $1 < X < 2$, $|X| > 1.5$, and $1 < |X| < 2$, respectively.

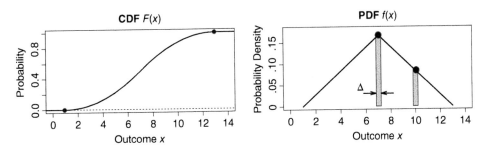

Figure 3.5 The CDF and PDF of an isosceles triangular distribution.

3.2.2 Using the PDF to Compute Relative Odds

For a discrete experiment, we may use the PMF $p(x)$ to compute the relative probability, or odds, that the outcome is $X = x_1$ versus $X = x_2$ as the ratio $p(x_1)/p(x_2)$. For the two-dice example shown in Table 3.3, the odds are 2:1 that we roll a 7 versus a 10, because $p(7) = \frac{1}{6}$ and $p(10) = \frac{1}{12}$. We can see this visually by comparing the heights of those values in the right frame of Figure 3.2 of the discrete isosceles triangular distribution.

However, in the corresponding continuous experiment shown in Figure 3.5, $f(7) = P(X = 7) = 0$ as does $f(10)$, and hence $\frac{0}{0}$ is not well defined. Yet, intuitively, the answer is 2:1 for this experiment as well.

The solution is to make small intervals of width Δ around 7 and 10:

$$\text{(odds of 7 to 10)} \approx \frac{P(X \approx 7)}{P(X \approx 10)}$$

$$= \frac{P\left[X \in \left(7 - \frac{\Delta}{2}, 7 + \frac{\Delta}{2}\right)\right]}{P\left[X \in \left(10 - \frac{\Delta}{2}, 10 + \frac{\Delta}{2}\right)\right]}.$$

Now for any continuous PDF, using the rectangle rule,

$$P(X \approx x) = \int_{x-\Delta/2}^{x+\Delta/2} f(x)\, dx \approx \Delta \cdot f(x) + o(\Delta),$$

where $o(\Delta)$ is defined in Appendix A. Therefore, for very small Δ,

$$\text{(odds of 7 to 10)} \xrightarrow{\Delta \to 0} \frac{\Delta \cdot f(7) + o(\Delta)}{\Delta \cdot f(10) + o(\Delta)} = \frac{f(7)}{f(10)},$$

which follows our intuition. In general, we have shown that the relative heights of the continuous PDF gives the relative odds:

$$\boxed{\textbf{odds } (X = x_1) \textbf{ versus } (X = x_2) = \frac{f(x_1)}{f(x_2)}, \quad \text{where } f \text{ is continuous.}} \qquad (3.11)$$

3.3 Summary Statistics: Central and Non-Central Moments

If we play a game of chance in which we must ante up a stake, then, on average, we may expect to win back some fraction of our bet. Suppose, for example, we pay \$1 for a chance to win \$27 if we roll snake eyes (two aces or ones). Then with probability 35/36 we win nothing, and with probability 1/36 we win \$27. So on average, if we play the game 36 times, we would win \$27, or 75 ¢ per game, which, given our stake, amounts to an average loss of 25 ¢ per game.

A game of chance where our expected gain is **zero** is called a **fair game**. Some games such as poker involve both chance and skill, and our expected gain might be positive (depending upon the skill of our opponents).

3.3.1 Expectation, Average, and Mean

In general, we call the average outcome of a random experiment characterized by a random variable X with CDF, $F(x)$, the **mean**, which we denote by the Greek letter μ and define by

$$\mu = E[X] = \begin{cases} \sum_x x \, p(x) & \text{if } X \text{ is discrete} \\ \int_{-\infty}^{\infty} x f(x) \, dx & \text{if } X \text{ is continuous.} \end{cases} \tag{3.12}$$

The symbol E is the expectation operation and averages over all possible outcomes, x, according to the probability law, $F(x)$. If we are interested in the average of some function, u, of X rather than X itself, the expectation is defined as

$$E[u(X)] = \begin{cases} \sum_x u(x) \, p(x) & \text{if } X \text{ is discrete} \\ \int_{-\infty}^{\infty} u(x) f(x) \, dx & \text{if } X \text{ is continuous.} \end{cases} \tag{3.13}$$

For example, $u(x)$ could be the winnings if the outcome is x. Notice how the random variable X on the left is replaced on the right by the particular values, x, weighted by their probability.

A useful special case of Equation (3.13) occurs when $u(\cdot)$ is the indicator function $I_A(\cdot)$ of the occurrence of an event of interest, A. Let

$$u(X) = I_A(X) = \begin{cases} 1 & \text{if } X \in A \\ 0 & \text{if } X \notin A. \end{cases}$$

Then $E[u(X)] = 1 \cdot P(X \in A) + 0 \cdot P(X \notin A)$ or

$$E[I_A(X)] = P(A), \quad \text{indicator functions.} \tag{3.14}$$

A more general version of Equation (3.14) is (note the use of $v(X)$, not $u(X)$)

$$E\{I[v(X) \in A]\} = P(v(X) \in A), \quad \text{indicator functions.} \tag{3.15}$$

Understanding notation matters for a good grasp of this material. For example, the expectations of $u(X)$ and $u(x)$ are not the same. The latter is not a random quantity (since using lower case x). The expectation of a constant is the constant itself.

3.3.2 Expectation as a Linear Operator

Since the summation and integral operators are linear, the expectation operator is linear as well; that is, for constants a and b,

$$E\left[a + b \cdot u(X)\right] = \int_{-\infty}^{\infty} \left[a + b \cdot u(x)\right] f(x)\, dx$$

$$= \int_{-\infty}^{\infty} a\, f(x)\, dx + \int_{-\infty}^{\infty} b \cdot u(x)\, f(x)\, dx$$

$$= a \int_{-\infty}^{\infty} f(x)\, dx + b \int_{-\infty}^{\infty} u(x)\, f(x)\, dx$$

$$\boxed{E\left[a + b \cdot u(X)\right] = a + b \cdot E[u(X)]\, ,} \tag{3.16}$$

using Equations (3.9) and (3.12). The same holds in the discrete case. In particular, we have from Equation (3.16)

$$E(a + X) = a + E(X)$$
$$E(bX) = b \cdot E(X).$$

We will use these expressions often in this book. We begin by introducing the variance in the next section.

3.3.3 The Variance of a Random Variable

This property of the expectation operator allows us to take a number of shortcuts in computation. Consider, for example, the variance of the r.v. X, var(X), which we denote by the Greek letter σ^2:

$$\sigma^2 = \text{var}(X) = E[(X - \mu)^2] = E[X^2 - 2\mu X + \mu^2]$$
$$= E[X^2] - E[2\mu X] + E[\mu^2]$$
$$= E[X^2] - 2\mu E[X] + \mu^2$$
$$= E[X^2] - 2\mu \cdot \mu + \mu^2 \qquad \text{or}$$

$$\boxed{\sigma^2 = \text{var}(X) = E[(X - \mu)^2] = E[X^2] - \mu^2}\ . \tag{3.17}$$

The variance summarizes the average squared distance of a r.v. from its mean. The following two properties of the variance will often be used in our calculations.

$$\boxed{(1) \quad \text{var}(a + X) = \text{var}(X)} \tag{3.18}$$

$$\boxed{(2) \quad \text{var}(b \cdot X) = b^2 \cdot \text{var}(X).} \tag{3.19}$$

The first relationship follows defining the r.v. $Y = a + X$. Now $EY = a + \mu_X$, where have added the subscript to distinguish μ_X from μ_Y. Hence,

$$\text{var}(Y) = E\left[(Y - \mu_Y)^2\right]$$
$$= E\left\{[(a + X) - (a + \mu_X)]^2\right\}$$
$$= E\left[(X - \mu_X)^2\right] = \text{var}(X).$$

Likewise, the second relationship follows defining the r.v. $T = b \cdot X$, for which $ET = b \cdot \mu_X$:

$$\begin{aligned}
\text{var}(T) &= E\left[(T - \mu_T)^2\right] \\
&= E\left[(b \cdot X - b \cdot \mu_X)^2\right] \\
&= E\left[b^2 \cdot (X - \mu_X)^2\right] \\
&= b^2 \cdot E\left[(X - \mu_X)^2\right] = b^2 \text{ var}(X).
\end{aligned}$$

These relationships hold for both discrete and continuous random variables. Later, when we study the units of random variables and expectation, we will find it convenient to scale and transform these quantities using

$$\boxed{\sigma = \sqrt{\sigma^2} = \sqrt{E[(X - \mu)^2]} \qquad \textbf{standard deviation,}} \tag{3.20}$$

which is the square root of the variance.

Example: The indicator function.   Since $I_A(X)^2 = I_A(X)$, then $E[I_A(X)^2] = P(A)$ as well. Hence,

$$\begin{aligned}
\text{var}(I_A(X)) &= E[I_A(X)^2] - \{E[I_A(X)]\}^2 \\
&= P(A) - P(A)^2 \\
&= P(A)\,[1 - P(A)]\ .
\end{aligned}$$

This variance is 0 if $P(A) = 0$ or 1 (why?), and is greatest when $P(A) = \frac{1}{2}$.

3.3.4   Standardized Random Variables

Suppose we consider a particular linear transformation of a random variable, X, which has finite mean μ_X and finite standard deviation, σ_X. The transformation is defined as

$$\boxed{Z = \frac{X - \mu_X}{\sigma_X}\ , \quad \text{standardized random variable.}} \tag{3.21}$$

Let us compute the first two moments of the new random variable, Z.

$$\begin{aligned}
\mu_Z = E[Z] &= E\left[\frac{X - \mu_X}{\sigma_X}\right] = \frac{1}{\sigma_X} E[X - \mu_X] \\
&= \frac{1}{\sigma_X}\left(E[X] - E[\mu_X]\right) = \frac{1}{\sigma_X}(\mu_X - \mu_X) = 0\ .
\end{aligned}$$

$$\begin{aligned}
\sigma_Z^2 = E[(Z - \mu_Z)^2] = E[Z^2] &= E\left[\left(\frac{X - \mu_X}{\sigma_X}\right)^2\right] \\
&= \frac{1}{\sigma_X^2} E[(X - \mu_X)^2] = \frac{1}{\sigma_X^2} \cdot \sigma_X^2 = 1\ .
\end{aligned}$$

Thus the mean of a standardized random variable Z is 0, while its variance and standard deviation are both 1.

3.3.5 Higher Order Moments

The kth order **non-central moment** and the kth order **central moment** of the random variable X are defined to be, respectively,

$$\mu'_k = E[X^k] \qquad \text{non-central moments} \tag{3.22}$$

$$\mu_k = E[(X - \mu)^k] \qquad \text{central moments.} \tag{3.23}$$

We have seen in Equation (3.17) how the central and non-central moments are related when $k = 2$.

The mean and the second–fourth central moments are commonly used and have their own special symbols. We list three here:

$$\mu_2 = \sigma^2 = E[(X - \mu)^2] \qquad \textbf{variance} \tag{3.24}$$

$$\mu_3 = \gamma = E[(X - \mu)^3] \qquad \textbf{skewness} \tag{3.25}$$

$$\mu_4 = \kappa = E[(X - \mu)^4] \qquad \textbf{kurtosis.} \tag{3.26}$$

Later when we study the units of random variables and expectation, we will find it convenient to scale and transform these quantities as follows: γ/σ^3 and κ/σ^4 as the standardized skewness and kurtosis, respectively. These are dimensionless quantities.

3.3.6 Moment Generating Function

Calculating all of these moments can become quite tedious. Fortunately, in this era of easy symbolic computation, there is an alternative. As the name suggests, the moment generating function (MGF), which is defined as

$$M_X(t) = E[e^{tX}] = \int_{-\infty}^{\infty} e^{tx} f(x) \, dx \,, \tag{3.27}$$

allows us to extract the kth non-central moment via differentiation, and finally the central moments via manipulations such as in Equation (3.17). Assuming the MGF is a "nice" function, then

$$\frac{d^k}{dt^k} M_X(t) = \int_{-\infty}^{\infty} \frac{d^k}{dt^k} \left[e^{tx} f(x) \right] \, dx$$

$$= \int_{-\infty}^{\infty} \left[\frac{d^k}{dt^k} e^{tx} \right] f(x) \, dx$$

$$= \int_{-\infty}^{\infty} x^k \, e^{tx} f(x) \, dx$$

$$= \int_{-\infty}^{\infty} x^k \, f(x) \, dx \quad \text{when } t = 0.$$

Thus the kth non-central moment, which we denote by μ'_k, is given by

$$\mu'_k = \frac{d^k}{dt^k} M_X(t) \bigg|_{t=0} . \tag{3.28}$$

Example: From Equation (3.27), Consider the uniform density on $(0, 1)$, for which $f(x) = 1$. The MGF of the Unif$(0, 1)$ density is

$$M_X(t) = \int_{x=0}^{1} e^{tx} \cdot 1 \, dx = \frac{e^t - 1}{t} .$$

Next, we compute $dM_X(t)/dt = (1 - e^t + te^t)/t^2$, which in the limit as $t \rightarrow 0$ equals $1/2$. Likewise, the second derivative equals $(-2 + 2e^t - 2te^t + t^2e^t)/t^3$, which equals $1/3$ in the limit as $t \rightarrow 0$. Thus, we have shown that $\mu_1' = \frac{1}{2}$ and $\mu_2' = \frac{1}{3}$; hence,

$$\mu = \mu_1' = \frac{1}{2} \quad \text{and} \quad \sigma^2 = \mu_2' - \mu^2 = \frac{1}{3} - \left[\frac{1}{2}\right]^2 = \frac{1}{12} ,$$

using Equation (3.17). Admittedly, it would have been much easier to compute these directly without the MGF, but for other examples forthcoming, the MGF has its advantages. In particular, the MGF has other important applications in analyzing sums of random variables.

Aside 1: Mathematica (Wolfram Research Inc. (2018)) was used to compute these formulae via

```
mgf = Integrate[ Exp[t x], {x,0,1} ]
Limit[ D[ mgf, {t,1} ], { t -> 0 } ]
Limit[ D[ mgf, {t,2} ], { t -> 0 } ].
```

Aside 2: The moment generating function is essentially the Laplace transform of the density $f(x)$. The Laplace transform is unique and invertible; however, it is not well defined for all values of t. A more general transformation is the **characteristic function** (CF), defined as the expectation of $\exp[itX]$, which exists for all values of t, where $i = \sqrt{-1}$. The CF is essentially the Fourier transform. We have no need for the extra generality, and will rely on the MGF only, rather than the CF.

3.3.7 Measurement Scales and Units of Measurement

The primary distinction we have drawn in the definition of a random variable is whether it is discrete or continuous. However, there is another characteristic of data measurement to which we should pay attention.

3.3.7.1 The Four Measurement Scales

The psychologist Stevens (1946) introduced four measurement scales that are useful in guiding appropriate statistical analysis. They are:

- **Nominal:** Categorical data that allows no possible meaningful order. Examples include eye color or types of pottery found at an archeological site.
- **Ordinal:** Data that are in an increasing (or decreasing) order. An example is a scale of strength of preference for movies.
- **Interval:** Data that are ordinal with the additional property that differences in the scale have the same meaning, no matter the location. An example is a date line on the Gregorian calendar.

- **Ratio:** Data that are interval and nonnegative with the additional property that there is a true zero. Thus concepts such as twice a quantity have real meaning. In the physical sciences, it is commonly assumed that all measurements follow a ratio scale, for example, mass and length.

Remarks: The three scales used in the measurement of temperature provide a nice example: Fahrenheit (°F), Celsius (°C), and Kelvin (K). All three are interval scales, but only Kelvin is a ratio scale. In the social sciences, development of scales that are interval (or nearly interval) from ordinal data is a matter of intense study in multivariate statistics. Standardized test scores are a good example of this ordinal/interval achievement and difficulty.

3.3.7.2 Units of Measurement

All continuous random variables have a unit of measurement, be it dollars for economic data or inches for height. How does the unit of measurement appear in our statistical models and summaries?

To be specific, let us assume the unit of measurement is dollars ($). If we assume the PDF is Unif(a, b), then the parameters a and b take on the unit of measurement dollars. Hence, the probability density of the uniform PDF has inverse units ($\$^{-1}$):

$$f(x|a, b) = \frac{1}{\$b - \$a} \qquad \$a \le x \le \$b,$$

since $b - a$ has units of $\$$ as well. We usually suppress the unit of measurement in our formulae. Note that a probability histogram (see Equation (1.2)) of economic data also has units of ($\$^{-1}$), as

$$\hat{f}(x) = \frac{v_k}{n \cdot h} = \frac{\text{count}}{\text{count} \cdot \$} = \frac{1}{\$},$$

since the bin width h is measured in $\$$.

If we consider the kth order sample non-central moment,

$$EX^k = \int_x x^k f(x) \, dx,$$

we see this has units $\$^k$, since the units of x^k are $\$^k$ while the units of the density ($\$^{-1}$) and dx ($\$$) cancel. Thus the units of the mean μ are in $\$$. The units of the variance (a central moment) are $\$^2$, as

$$\sigma^2 = E[(X - \mu)^2] = \int_x (x - \mu)^2 f(x) \, dx = (\$ - \$)^2 \cdot \frac{1}{\$} \cdot \$ = \$^2 .$$

The units of the standard deviation $\sigma = \sqrt{\sigma^2}$ are $\sqrt{\$^2}$ or just dollars $\$$.

3.4 Binomial Experiments

Next we develop the binomial distribution, which has widespread applicability. Each outcome of the random variable, X_1, is binary, and may be thought of as a success (S) or failure

(F), a head (H) or tail (T), or as with indicator functions, as a true (T) or false (F). We label the outcome as $x = 1$ or $x = 0$, respectively, for any of these interpretations. Rather than assume equally likely outcomes in this oft-called **Bernoulli trial**, we suppose that $p = P(X_1 = 1)$ is a (known) real number between 0 and 1. We do not need to manufacture a biased coin for each value of p; rather, we can use our spinner to generate a single outcome. The **R** function to simulate the result of a spinner is `runif(1)`. If the spinner lands in the interval $(0, p)$, then we have a success; if the spinner lands in the interval $(p, 1)$, then we have a failure. This is another example of a geometric probability argument, as the lengths of the two intervals are p and $1 - p$, respectively. We will denote the failure probability by $\boxed{q = 1 - p}$. Note that $P(X_1 = 0) = 1 - p$, the complement of the success probability. This uses Kolmogorov's third axiom, Equation (2.23). To generate true/false outcomes using **R**, type `runif(1)<p`.

The binomial distribution repeats this simple experiment, X_1, n times, and the random variable X counts the total number of successes. We assume that the outcome of the ith experiment is not affected by the knowledge of the previous experiments, that is, they are independent as in Equation (2.14). Let the random variable X_i measure the outcome of the ith experiment. Then

$$X = \sum_{i=1}^{n} X_i \quad \text{and we write } X \sim \text{Binom}(n, p) , \tag{3.29}$$

which means that the random variable X follows the one-parameter binomial distribution with parameter p. Next, we attempt to derive the PMF of X.

Focusing on the biased coin example, a typical sequence of heads and tails might be (note that this sequence uses our intersection notational shorthand)

$$H_1 T_2 T_3 T_4 \cdots H_{n-2} T_{n-1} H_n . \tag{3.30}$$

Since each Bernoulli trial is independent, the probability of this sequence is

$$p \times (1 - p) \times (1 - p) \times (1 - p) \times \cdots \times p \times (1 - p) \times p = p^x \, q^{n-x} ,$$

where x is the number of heads, and $n - x$ is the number of tails.

The sequence in Equation (3.30) is only one of many where $X = x$. To count them all, we note that we can focus on the locations of the x heads in the n positions. Clearly, order does not matter in this process. Thus, there are $\binom{n}{x}$ ways of selecting the positions. For each such selection, there is only 1 way for the $n - x$ tails to complete the list. If we instead focused on placing the $n - x$ tails in the list, there would be $\binom{n}{n-x}$ ways of selecting the positions. Fortunately, both approaches give the same answer since $\binom{n}{x} = \binom{n}{n-x}$. Thus we have derived the Binom(n, p) PMF:

$$\boxed{p(x) = P(X = x) = \binom{n}{x} p^x q^{n-x}, \quad x = 0, 1, \ldots, n \quad \text{Binom}(n, p).} \tag{3.31}$$

That this is indeed a valid probability distribution follows from an application of the binomial expansion given in Appendix A for polynomials:

$$\sum_{x=0}^{n} p(x) = \sum_{x=0}^{n} \binom{n}{x} p^x q^{n-x} = (p + q)^n = 1^n = 1 .$$

The **R** functions dbinom, pbinom, qbinom, and rbinom compute the PDF, CDF, percentiles, and random samples for the Binom(n, p) distribution. Note that the PMF $p(x)$ also uses p, but as a function, not the binomial probability parameter. Hopefully this confusion will be only momentary.

We compute the moments through the MGF rather than directly:

$$M_X(t) = E[e^{tX}] = \sum_{x=0}^{n} e^{tx} \cdot \binom{n}{x} p^x q^{n-x}$$

$$= \sum_{x=0}^{n} \binom{n}{x} (pe^t)^x q^{n-x} \qquad \text{or}$$

$$\boxed{M_X(t) = (pe^t + q)^n , \qquad X \sim \text{Binom}(n, p) .} \tag{3.32}$$

Using Equation (3.28), we have $\mu_1' = np$ and $\mu_2' = np + n(n-1)p^2$; hence,

$$\boxed{\mu = \mu_1' = np} \qquad \text{and} \tag{3.33}$$

$$\sigma^2 = \mu_2' - \mu^2 = np + n(n-1)p^2 - (np)^2 = np - np^2 \quad \text{or}$$

$$\boxed{\sigma^2 = np(1-p) = npq ; \quad \text{hence, } \sigma = \sqrt{npq} .} \tag{3.34}$$

In Figure 3.6, we display the (discrete) binomial PMF for five choices of p and three choices of n. The corresponding CDFs are displayed in Figure 3.7. Note for the larger values of n, the PMF is centered (nearly symmetrically) around the mean $\mu = np$ and that the shape of the PMF looks like the bell-shaped normal curve for all five choices of p.

Example: A student applying for security clearance needs to score at least 80% on a true/-false test with 25 questions. If she is forced to guess at all 25 questions, what is the probability that she passes? What if she knows the answers for 15 of the questions, and guesses at the rest? (She can miss 5.)

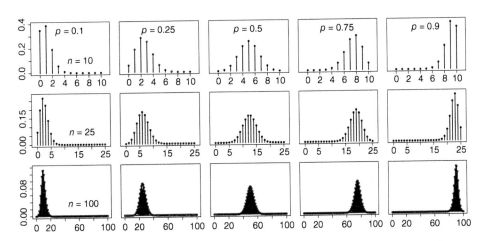

Figure 3.6 Binomial PMF for various combinations of n and p.

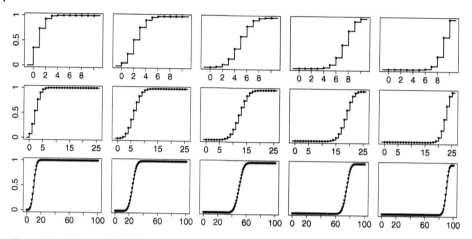

Figure 3.7 Binomial CDF for n and p as in Figure 3.6.

Answers: The probabilities of passing are $\sum_{x=20}^{25} p(x|n = 25, p = 0.5)$ and $\sum_{x=5}^{10} p(x|n = 10, p = 0.5)$, respectively. Here, we used the **R** commands

```
sum( dbinom( 20:25, 25, 0.5) ) = 0.2%
sum( dbinom( 5:10, 10, 0.5) ) = 62.3% ;
```

she had better study. Examine the middle columns in Figures 3.6 and 3.7.

3.5 Waiting Time for a Success: Geometric PMF

If we perform a series of Binom$(1, p)$ Bernoulli trials, how long do we need to wait until we observe the first success? Let X denote the number of trials until the first success is observed. Clearly, X can be any positive integer. If $X = x$ is observed, then the pattern of success and failures must have been

$$
\begin{array}{c|ccccccc}
\text{outcome} & F & F & F & \cdots & F & F & S \\
\hline
\text{trial number} & 1 & 2 & 3 & & x-2 & x-1 & x
\end{array}
\qquad (3.35)
$$

which has probability $q^{x-1}p = p(x)$, $x = 1, 2, \ldots, \infty$. Note that

$$
\sum_{x=1}^{\infty} p(x) = \sum_{x=1}^{\infty} q^{x-1}p = \frac{p}{q}\sum_{x=1}^{\infty} q^x = \frac{p}{q}\cdot\frac{q}{1-q} = 1 \quad \text{and}
$$

$$
M_X(t) = \sum_{x=0}^{\infty} e^{tx}\cdot q^{x-1}p = \frac{p}{q}\sum_{x=1}^{\infty} e^{tx}\cdot q^x
$$

$$
= \frac{p}{q}\sum_{x=1}^{\infty}(qe^t)^x = \frac{p}{q}\cdot\frac{qe^t}{1-qe^t} = \frac{pe^t}{1-qe^t} ,
$$

if $|qe^t| < 1$. Using Mathematica, $\mu'_1 = 1/p$ and $\mu'_2 = (2-p)/p^2$; hence, $\mu = \mu'_1 = 1/p$ and $\sigma^2 = \mu'_2 - \mu^2 = q/p^2$.

Example: Near the end of a hard fought backgammon game, we have two pieces on the bar waiting to re-enter the opponent's inner board, but the opponent has covered points 2–6. Only the first point is open. We need to roll snake eyes to have any hope, and we guess we have five turns before our opponent covers the 1 spot as well. What is the probability we can get both of our pieces off the bar?

Answer: The probability is $P(X \le 5) = F_X(5) = \sum_{x=1}^{5} q^{x-1}p = 13.1\%$ with $p = 1/36$. We should probably resign if offered the doubling cube.

3.6 Waiting Time for r Successes: Negative Binomial

Suppose we ask how many **failures**, X, do we observe before we observe the rth success; then we have the so-called **negative binomial** distribution. The random variable $X \sim \text{NegBinom}(r, p)$ can take on any value greater than or equal to zero, with the experiment terminating on trial number $r + x$. The sequence of failures and successes is similar to what we have seen in Equation (3.35) ending with a success, but with $r - 1$ successes in the first $r + x - 1$ places, for example, 10 trials 0001001001 to obtain $r = 3$ successes and $x = 7$ failures. Each sequence has probability $p^r q^x$. The number of sequences with x fixed is multiplied by $\binom{x+r-1}{x}$, which counts the number of positions where the x failures may fall in the first $x + r - 1$ positions. Hence,

$$p(x) = \binom{r + x - 1}{x} p^r q^x , \quad x = 0, 1, \ldots \infty \quad \textbf{negative binomial.} \tag{3.36}$$

Remark: Since it can be shown that $\mu = r \times p/q$, a salesman with a quota to sell r items will have to endure a number of unsuccessful sales pitches. The **R** function dnbinom may be used to compute specific probabilities.

3.7 Poisson Process and Distribution

As the telephone system expanded rapidly in the early 20th century, Bell System scientists modeled the demand on their switchboard operators. The Poisson process is very useful in this regard, showing how different mathematical tools can aid in the derivation of a new PMF.

Let the random variable X count the number of calls in the time interval $[0, t]$. We make only three assumptions:

1. The rate, λ, at which calls are received is constant.
2. Calls in non-overlapping time intervals are independent.
3. The probability of a call in a small time interval $[0, \delta]$ is proportional to δ.

Consider the third assumption. If δ is sufficiently small, then most of the time there will be 0 calls, but with small probability there might be 1 call (and a vanishing probability

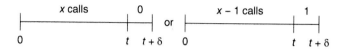

Figure 3.8   The two disjoint events that result in *x* calls in $[0, t + \delta]$, ignoring the very small possibility of more than one call in $(t, t + \delta)$.

of more than 1 call). This is approximately a Binom$(1, p)$ event, so $\mu = p$. Now from the first assumption, we expect there to be $\lambda\delta << 1$ calls in the short time interval. (Why?) More precisely, we can express the third assumption as $p = \lambda\delta + o(\delta)$. To derive the Poisson PMF, we examine how *x* calls could arrive in the slightly wider time interval $[0, t + \delta]$ by counting calls in the two disjoint intervals $[0, t]$ and $(t, t + \delta]$; see Figure 3.8. These calls are independent by the second assumption.

Define $P(x, t) =$ probability exactly *x* calls are received in time *t*. Then ignoring terms of order $o(\delta)$,

$$P(x, t + \delta) = P(x, t) \cdot P(0, \delta) + P(x - 1, t) \cdot P(1, \delta)$$

$$= P(x, t) \cdot (1 - \lambda\delta) + P(x - 1, t) \cdot \lambda\delta ,$$

where the probabilities add because the possibilities of the event sequences $(x, 0)$ and $(x - 1, 1)$ in Figure 3.8 are disjoint; furthermore, the probabilities within each event multiply because the time intervals $[0, t]$ and $(t, t + \delta]$ are non-overlapping and the counts are independent by the second assumption. Rearranging,

$$\frac{P(x, t + \delta) - P(x, t)}{\delta} = -\lambda P(x, t) + \lambda P(x - 1, t) \qquad \text{or}$$

$$\text{letting } \delta \to 0, \qquad \frac{\partial P(x, t)}{\partial t} = -\lambda P(x, t) + \lambda P(x - 1, t) .$$

It is straightforward to check that the solution is

$$\boxed{P(x, t) = \frac{e^{-\lambda t}(\lambda t)^x}{x!}} , \qquad (3.37)$$

because using the product rule gives

$$\frac{\partial P(x, t)}{\partial t} = \frac{-\lambda e^{-\lambda t}(\lambda t)^x}{x!} + \frac{e^{-\lambda t}x (\lambda t)^{x-1}\lambda}{x!}$$

$$= -\lambda P(x, t) + \lambda \frac{e^{-\lambda t}(\lambda t)^{x-1}}{(x - 1)!}$$

$$= -\lambda P(x, t) + \lambda P(x - 1, t) .$$

3.7.1   Moments of the Poisson PMF

A common parameterization of the Poisson PMF is

$$\boxed{X \sim \text{Pois}(m) \qquad \text{where} \qquad p(x) = \frac{e^{-m} m^x}{x!} \qquad x = 0, 1, \dots, \infty .} \qquad (3.38)$$

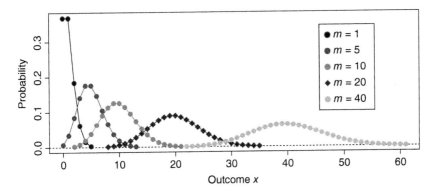

Figure 3.9 Examples of the Poisson PMF, where $X \sim \text{Pois}(m)$.

Note

$$\sum_{x=0}^{\infty} p(x) = e^{-m} \sum_{x=0}^{\infty} \frac{m^x}{x!} = e^{-m} \cdot e^m = 1 .$$

$$M_X(t) = \sum_{x=0}^{\infty} e^{tx} \cdot \frac{e^{-m} m^x}{x!} = e^{-m} \sum_{x=0}^{\infty} \frac{e^{tx} m^x}{x!}$$

$$= e^{-m} \sum_{x=0}^{\infty} \frac{(me^t)^x}{x!} = e^{-m} \cdot e^{me^t} ; \qquad \text{hence}$$

$$M_X(t) = e^{m(e^t-1)} . \tag{3.39}$$

Using Mathematica, $\mu'_1 = m$ and $\mu'_2 = m(1-m) = m - m^2$; hence,

$$X \sim \text{Pois}(m) \qquad \Rightarrow \qquad \mu = m , \qquad \sigma^2 = m . \tag{3.40}$$

3.7.2 Examples

Question: The Meyerland subdivision of Houston, Texas, has had three 100-year storms during the past decade ($t = 10$ years). How likely is that?

Answer: Since $\lambda = 1$ storm/100 years, then $m = \lambda t = 0.01 \times 10 = 0.1$. Then $P(X \geq 3) = 1 - P(X \leq 2) = 0.000155$, using the **R** function dpois. Highly unusual. Of course, storms in adjacent years may not be independent, and the Poisson model may not be the best approximation in that case.

Some examples of the Poisson PMF are shown in Figure 3.9. Note the density also has the bell-shaped shape for large m.

3.8 Waiting Time for Poisson Events: Negative Exponential PDF

The negative exponential PDF is a continuous version of the geometric waiting time distribution discussed in Section 3.5. Let the random variable, T, denote the time to the next

Poisson event, using the definition in Equation (3.37). Again, $X \sim \text{Pois}(m = \lambda t)$. We make use of the fact that two events that always occur together (or alternatively never occur together) must have the same probability. In this setting, having to wait more than time t for the next event is equivalent to observing no events in $(0, t)$. Hence,

$$F_T(t) = P(T \leq t) = 1 - P(T > t) = 1 - P(X = 0)$$

$$\boxed{F_T(t) = 1 - \frac{e^{-\lambda t}(\lambda t)^0}{0!} = 1 - e^{-\lambda t}, \quad t \geq 0 .} \tag{3.41}$$

$$\boxed{f_T(t) = \frac{d \, F_T(t)}{d \, t} = \lambda e^{-\lambda t}, \quad t \geq 0 .} \tag{3.42}$$

We note

$$\int_{-\infty}^{\infty} f(t) \, dt = \int_0^{\infty} \lambda e^{-\lambda t} \, dt = 1 .$$

$$M_T(s) = \int_0^{\infty} e^{st} \cdot \lambda e^{-\lambda t} \, dt = \frac{\lambda}{\lambda - s} ,$$

where we have used the symbol s in the MGF to avoid confusion with the time variable, t. Using Mathematica, we find $\mu_1' = 1/\lambda$ and $\mu_2' = 2/\lambda^2$. Hence,

$$\mu = \frac{1}{\lambda} \quad \text{and} \quad \sigma^2 = \mu_2' - \mu^2 = \frac{1}{\lambda^2} .$$

The formula for the mean is satisfying, as the reciprocal of the average rate is the average period.

3.9 The Normal Distribution (Also Known as the Gaussian Distribution)

The normal density may be the most important distribution in both theory and practice, but there is no simple axiomatic derivation. Instead, there are various approximations and limiting arguments that lead to the normal distribution. Here, we take a simplistic algebraic approach. We note that the exponential distribution in the previous section is the exponential of a linear polynomial. Thus we consider the generalization to the exponential of a quadratic polynomial in x, namely,

$$f(x) = e^{-ax^2 + bx + c} = e^c \cdot e^{-ax^2 + bx} , \quad \text{where } a > 0 \text{ and } x \in \mathbb{R}^1 . \tag{3.43}$$

The constant a must be positive, or the integral of $f(x)$ would be infinite. The constant c is selected so that the area is 1. Computing the area using Mathematica gives

$$\int_{-\infty}^{\infty} f(x) \, dx = \sqrt{\frac{\pi}{a}} \exp\left[c + \frac{b^2}{4a}\right] = 1 \quad \Rightarrow \quad c = \log\sqrt{\frac{a}{\pi}} + \frac{-b^2}{4a} .$$

Substituting c into the density formula (3.43), we obtain

$$f(x) = \exp\left[\log\sqrt{\frac{a}{\pi}} + \frac{-b^2}{4a}\right] \cdot \exp[-ax^2 + bx]$$

$$= \sqrt{\frac{a}{\pi}} \, \exp\left[-\frac{b^2}{4a} - ax^2 + bx\right]$$

$$= \sqrt{\frac{a}{\pi}} \, \exp\left[-a\left(x^2 - \frac{b}{a}x + \frac{b^2}{4a^2}\right)\right] = \sqrt{\frac{a}{\pi}} \, \exp\left[-a\left(x - \frac{b}{2a}\right)^2\right].$$

Next, we directly compute the mean, $\mu = \int x f(x) \, dx = b/2a$ and the variance $\sigma^2 = \int (x - \mu)^2 f(x) \, dx = 1/2a$, so that $a = 1/2\sigma^2$. Thus the three parameters a, b, and c can be rewritten in terms of μ and σ^2, giving

$$X \sim N(\mu, \sigma^2) \quad \Longleftrightarrow \quad f(x) = \frac{1}{\sqrt{2\pi\sigma^2}} \exp\left[-\frac{1}{2\sigma^2}(x - \mu)^2\right]. \tag{3.44}$$

It is very convenient to write the normal density in terms of the first two moments, rather than the equally correct form in Equation (3.43).

The **standard normal** PDF has $\mu = 0$ and $\sigma^2 = 1$, and is usually denoted by the r.v. Z. The PDF is often denoted by ϕ; hence,

$$Z \sim N(0, 1) \quad \Longleftrightarrow \quad \phi(z) = \frac{1}{\sqrt{2\pi}} \, e^{-z^2/2}. \tag{3.45}$$

Before we derive any theoretical results, we can observe some of the approximation properties empirically. For example, the Poisson PMF Pois(m) and normal PDF with matching moments are quite close for large m, even though one is discrete and the other is continuous; see Figure 3.10. Recall $\mu = \sigma^2 = m$ for the Pois(m) PMF.

The MGF of the $N(\mu, \sigma^2)$ distribution is (using Mathematica)

$$X \sim N(\mu, \sigma^2) \quad \Longrightarrow \quad M_X(t) = \exp\left(\mu\, t + \frac{1}{2}\sigma^2\, t^2\right). \tag{3.46}$$

Example: If the moment generating function of a random variable X is $\exp[10t + 200t^2]$, then we may conclude that $X \sim N(10, 400)$, exactly.

Historical note: The Marquis de Laplace had a significant role in the early development of the normal distribution, proving the central limit theorem, for example. Abraham de Moivre was the first to encounter something like the normal distribution as an

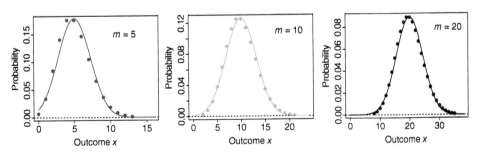

Figure 3.10 Examples of the discrete Poisson PMF, Pois(m), and the continuous normal PDF with the same moments, $N(\mu = m, \sigma^2 = m)$.

Figure 3.11 Gauss on the German Mark bill. Note the Gaussian curve.

approximation in the binomial theorem. However, Carl Friedrich Gauss used the normal density to motivate least squares, which may have given him naming rights. Germany honored Gauss in 1991, placing his image on its 10 Mark bank note; see Figure 3.11.

3.9.1 Standard Normal Distribution

The PDF of the random variable $X \sim N(\mu, \sigma^2)$ is easy to write down, but its CDF is not integrable in closed form. Thus tables are necessary. In lieu of tables, we use the **R** function `pnorm(x, μ = 0, σ = 1)`. (*Caution:* The **R** function `pnorm` takes σ, not σ^2, as its third argument.) It is common to use the Greek letters ϕ and Φ to denote the PDF and CDF of the **standard normal distribution**, $N(0, 1)$, respectively, unless μ and σ are otherwise included. See Figure 3.12 for an example finding $\Phi(x = 1)$ when $X \sim N(0, 1)$. The probability is also given by the shaded area under the PDF, since $\Phi(1) = \int_{x=-\infty}^{1} \phi(x) \, dx$.

Again, it is common to use the symbol Z when standardizing a normal $N(\mu, \sigma^2)$ r.v. From Section 3.3.4, we know the mean and variance of

$$Z = \frac{X - \mu}{\sigma} \tag{3.47}$$

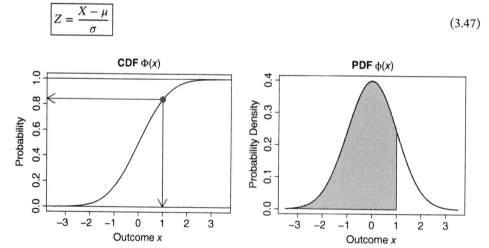

Figure 3.12 Standard normal CDF, $\Phi(x)$, and PDF, $\phi(x)$, for $x = 1$.

are always 0 and 1, respectively, but what is the distribution of Z? To answer that question, we compute the MGF of Z and see if we "recognize" it from our limited list (so far). Computing

$$M_Z(t) = E[\exp(tZ)] = E\left[\exp\left(t \cdot \frac{X - \mu}{\sigma}\right)\right]$$

$$= E\left[\exp\left(\frac{t}{\sigma} \cdot X - \frac{\mu t}{\sigma}\right)\right] = E\left[\exp\left(\frac{t}{\sigma} \cdot X\right) \cdot \exp\left(\frac{-\mu t}{\sigma}\right)\right]$$

$$= \exp\left(\frac{-\mu t}{\sigma}\right) \cdot E\left[\exp\left(\frac{t}{\sigma} \cdot X\right)\right]$$

$$= \exp\left(\frac{-\mu t}{\sigma}\right) \cdot \exp\left(\mu \cdot \left(\frac{t}{\sigma}\right) + \frac{1}{2}\sigma^2\left(\frac{t}{\sigma}\right)^2\right)$$

$$= \exp\left(\frac{1}{2}t^2\right) ; \tag{3.48}$$

hence from Equation (3.40), we can conclude that $Z \sim N(0, 1)$ exactly. Thus most textbooks that publish tables do so only for the standard normal distribution. The following identity makes this sufficient. If $X \sim N(\mu, \sigma^2)$,

$$P(X < c) = P\left(\frac{X - \mu}{\sigma} < \frac{c - \mu}{\sigma}\right) = P\left(Z < \frac{c - \mu}{\sigma}\right) \text{ or}$$

$$\boxed{P(X < c) = \Phi\left(\frac{c - \mu}{\sigma}\right).} \tag{3.49}$$

3.9.2 Sums of Independent Normal Random Variables

The MGF technique exemplified in the previous section can be extended to prove the following (exact) result. If the random variables $X_i \sim N(\mu_i, \sigma_i^2)$ and are independent then,

$$\boxed{Y = \sum_{i=1}^{n} a_i X_i \sim N\left(\sum_{i=1}^{n} a_i \mu_i, \sum_{i=1}^{n} a_i^2 \sigma_i^2\right).} \tag{3.50}$$

The proof is an exercise; see Problem 5.

Example: $\overline{X}$ is of this form if we take $a_i = 1/n$, $\mu_i = \mu$, and $\sigma_i = \sigma$. Then

$$\boxed{\overline{X} = \sum_{i=1}^{n} \frac{1}{n} X_i \sim N\left(\sum_{i=1}^{n} \frac{1}{n} \mu, \sum_{i=1}^{n} \left[\frac{1}{n}\right]^2 \sigma^2\right) \sim N\left(\mu, \frac{\sigma^2}{n}\right).} \tag{3.51}$$

3.9.3 Normal Approximation to the Poisson Distribution

We have seen in Figure 3.10 that the Poisson density looks very similar to a normal density as $m \to \infty$. In this section, we use the MGF to prove why this is true. First, we standardize the Poisson random variable, $X \sim \text{Pois}(m)$, by defining

$$Y = \frac{X - \mu}{\sigma} = \frac{X - m}{\sqrt{m}} .$$

We attempt to compute the MGF of Y and "recognize" it as before.

$$M_Y(t) = E[\exp(tY)] = E\left[\exp\left(t \cdot \frac{X - m}{\sqrt{m}}\right)\right]$$

$$= E\left[\exp\left(\frac{t}{\sqrt{m}} \cdot X - \sqrt{m}\, t\right)\right]$$

$$= E\left[\exp\left(\frac{t}{\sqrt{m}} \cdot X\right)\right] \cdot \exp(-\sqrt{m}\, t)$$

$$= \exp\left\{m\left[\exp\left(\frac{t}{\sqrt{m}}\right) - 1\right]\right\} \cdot \exp(-\sqrt{m}\, t)$$

$$= \exp\left\{m\left[\exp\left(\frac{t}{\sqrt{m}}\right) - 1\right] - \sqrt{m}\, t\right\}, \tag{3.52}$$

using formula (3.39), replacing t with $t/\sqrt{m}$. Next, we use Mathematica to find a Taylor series for the quantity in curly brackets in Equation (3.52), `Series[ %1, {m, Infinity, 1}]`, obtaining

$$M_Y(t) = \exp\left\{\frac{1}{2}t^2 + \frac{1}{6}\frac{t^3}{\sqrt{m}} + \frac{1}{24}\frac{t^4}{m} + \cdots\right\}. \tag{3.53}$$

As $m \to \infty$, the MGF converges to $\exp(t^2/2)$, which we recognize as the MGF of the standard $N(0, 1)$ distribution, just as we did in Equation (3.48).

Remarks: This result is a special case of the **central limit theorem (CLT)**, which we will study in Section 5.4. Loosely speaking, we may write our finding as

$$\boxed{X \sim \text{Pois}(m) \approx N(m, m)\,, \qquad \text{for large } m\,.} \tag{3.54}$$

Problems

3.1 Throw three dice and label the results X_1, X_2, and X_3. What is the PMF of the total number of pips, $X = X_1 + X_2 + X_3$?

3.2 Find the skewness and kurtosis coefficients in terms of μ and the non-central moments μ'_2, μ'_3, and μ'_4.

3.3 If $X \sim \text{NegBinom}(r, p)$, find its MGF; then its mean and variance.

3.4 Use Mathematica to find the Taylor series approximation to the difference of the exact MGF in Equation (3.52) and the MGF of a standard normal. (We used a Taylor series in the exponent when deriving Equation (3.53).)

3.5 Use the MGF technique to prove the result in Equation (3.50).
Hint: You should use results from Section 5.4.1

4

Bivariate Random Variables, Transformations, and Simulations

We have now completed our initial exploratory of the distribution and shape of data. The CDF and associated PMF or PDF completely characterize this knowledge. In this chapter, we explore the ways in which bivariate data interact, and how that interaction can used for purposes of prediction. Our graphical exploration of the Pearson father–son data hints at what our bivariate models will capture analytically.

4.1 Bivariate Continuous Random Variables

If two measurements are made after an experiment is run, then we represent the result in the vector (X, Y) or perhaps as (X_1, X_2). There is no limit on the number of measurements that can be made, for example, all 20,000 genes. Again, we assume the measurements are made perfectly. There is no observation noise.

4.1.1 Joint CDF and PDF Functions

The bivariate cumulative distribution function of (X, Y) is defined to be

$$F_{X,Y}(x,y) = P(X \le x, Y \le y) \quad \text{for } (x,y) \in \mathbb{R}^2. \tag{4.1}$$

Clearly, the bivariate CDF, $F_{X,Y}$, is a monotone non-decreasing function satisfying $F_{X,Y}(-\infty, -\infty) = 0$ while $F_{X,Y}(\infty, \infty) = 1$. By adding and subtracting values of the CDF, we can find the probabilities associated with events in simple geometric shapes, such as rectangles and squares.

Generalizing Equations (3.5) and (3.6), the bivariate probability density function is defined to be

$$f_{X,Y}(x,y) = \frac{\partial^2 F_{X,Y}(x,y)}{\partial x \, \partial y} \ge 0. \tag{4.2}$$

Knowing the PDF allows us to recover the CDF:

$$F_{X,Y}(x,y) = \int_{t=-\infty}^{y} \int_{s=-\infty}^{x} f_{X,Y}(s,t) \, ds \, dt. \tag{4.3}$$

Statistics: A Concise Mathematical Introduction for Students, Scientists, and Engineers, First Edition. David W. Scott.
© 2020 John Wiley & Sons Ltd. Published 2020 by John Wiley & Sons Ltd.

Finally, the extension of Equation (3.10) to two dimensions is

$$P((X, Y) \in A) = \iint\limits_{(x,y) \in A} f_{X,Y}(x, y) \, dx \, dy.$$ (4.4)

4.1.2 Marginal PDF

The joint CDF $F_{X,Y}(x, y)$ and PDF $f_{X,Y}(x, y)$ contain all the information about the univariate distributions of X and Y as well. Note the subscripts to distinguish the CDF $F_X(x)$ and the PDF $f_X(x)$. In particular,

$$F_X(x) = P(X \le x, Y \le \infty) = F_{X,Y}(x, \infty)$$

$$= \int_{t=-\infty}^{\infty} \left[\int_{s=-\infty}^{x} f_{X,Y}(s, t) \, ds \right] dt.$$

Recall the second fundamental theorem of calculus:

$$\frac{d}{dx} \int_a^x g(t) \, dt = g(x).$$

Hence,

$$f_X(x) = \frac{d}{dx} F_X(x) = \int_{t=-\infty}^{\infty} \left[\frac{d}{dx} \int_{s=-\infty}^{x} f_{X,Y}(s, t) \, ds \right] dt \quad \text{or}$$

$$f_X(x) = \int_{t=-\infty}^{\infty} [f_{X,Y}(x, t)] \, dt, \quad \text{or, equivalently}$$

$$f_X(x) = \int_{y=-\infty}^{\infty} f_{X,Y}(x, y) \, dy, \quad \text{the marginal PDF of } X.$$ (4.5)

Informally, one speaks of "integrating out" the other variable. Similarly,

$$f_Y(y) = \int_{x=-\infty}^{\infty} f_{X,Y}(x, y) \, dx, \quad \text{the marginal PDF of } Y.$$ (4.6)

For bivariate discrete PMFs, finding the marginals is only a matter of bookkeeping. In Figure 4.1, we see that the marginals of X and Y are identical: $p_X(x) = p_Y(y) = \left\{ \frac{4}{10}, \frac{3}{10}, \frac{2}{10}, \frac{1}{10} \right\}$ for x and y equal to 1, 2, 3, 4. In the discrete case, we "sum out" the other variable.

4.1.3 Conditional Probability Density Function

In the bivariate setting (X, Y), if we know that $X = x_0$, then the slice of the bivariate density given by $f_{X,Y}(x_0, y)$ is proportional to the desired conditional PDF $f_{Y|X=x_0}(y|x_0)$ by the relative odds argument. To make this a PDF, we need to renormalize so the integral of the conditional density is one:

$$f_{Y|X=x_0}(y|x_0) = \frac{f_{X,Y}(x_0, y)}{f_X(x_0)}, \quad \textbf{conditional PDF.}$$ (4.7)

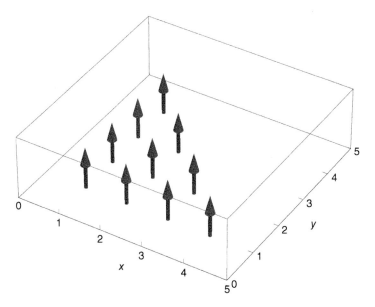

Figure 4.1 Joint bivariate PMF. Each arrow displays a probability of $\frac{1}{10}$.

Note by Equation (4.5), $\int_y f_{X,Y}(x_0, y)\, dt = f_X(x_0)$; therefore, the integral of Equation (4.7) over y is one.

The continuous form of the Bayes theorem (2.15) is given by

$$\boxed{f_{Y|X=x}(y|x) = \frac{f_{X|Y=y}(x|y)f_Y(y)}{f_X(x)}, \quad \textbf{continuous r.v. Bayes theorem}.} \tag{4.8}$$

Example: Consider the bivariate PDF over $\mathbb{R}^2$ given by

$$f_{X,Y}(x,y) = \frac{5}{6\pi}\, e^{-\frac{25}{18}\left(x^2 - \frac{8}{5}xy + y^2\right)}, \tag{4.9}$$

which is shown in the left frame of Figure 4.2. Suppose we know that $X = 1$. Then the conditional density is proportional to the function $f_{X,Y}(x,y)$ along y at $x = 1$. This is depicted

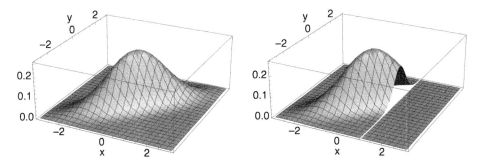

Figure 4.2 Conditional PDF, $f_{Y|X=1}(y|1)$, before normalization.

in the right frame of Figure 4.2. Explicitly,

$$f_{Y|X=1}(y|x=1) \propto f_{X,Y}(1,y) \propto e^{-\frac{25}{18}\left(1-\frac{8}{5}y+y^2\right)}, \tag{4.10}$$

where the symbol "$\propto$" means proportional to. Using Mathematica, we find the marginal PDF of Equation (4.9) is $f_X(x) = N(0,1)$, and that the integral of Equation (4.10) is $e^{-1/2}3\sqrt{2\pi}/5$; hence, a little algebra shows

$$f_{Y|X=1}(y|x=1) = \frac{5}{3\sqrt{2\pi}} e^{-\frac{25}{18}\left(y-\frac{4}{5}\right)^2} \sim N\left(\frac{4}{5}, \frac{9}{25}\right). \tag{4.11}$$

Reproducing for arbitrary $X = x$, we find the conditional PDF is given by

$$\boxed{f_{Y|X=x}(y|x) \sim N\left(\frac{4}{5}x, \frac{9}{25}\right).} \tag{4.12}$$

4.1.4 Independence of Two Random Variables

Let the CDFs of the r.v.s X and Y be $F_X(x)$ and $F_Y(y)$, respectively. If X and Y are independent, then the bivariate CDF factors:

$$\begin{aligned} F_{X,Y}(x,y) &= P(X \leq x \text{ and } Y \leq y) \\ &= P(X \leq x) \cdot P(Y \leq y) \\ &= F_X(x) \cdot F_Y(y). \end{aligned}$$

Therefore, the bivariate probability density becomes

$$f_{X,Y}(x,y) = \frac{\partial^2 F_{X,Y}(x,y)}{\partial x \, \partial y} = \frac{\partial^2 [F_X(x)F_Y(y)]}{\partial x \, \partial y} = \frac{\partial F_X(x)}{\partial x} \cdot \frac{\partial F_Y(y)}{\partial y}$$

$$\boxed{f_{X,Y}(x,y) = f_X(x) \cdot f_Y(y), \qquad \textbf{definition of independence.}} \tag{4.13}$$

We take Equation (4.13) as defining when two continuous random variables are independent.

4.1.5 Expectation, Correlation, and Regression

For bivariate continuous random variables (X, Y) with joint PDF $f_{X,Y}(x,y)$, expectation is generally defined as

$$\boxed{E[g(X, Y)] = \int_y \int_x g(x,y) \, f_{X,Y}(x,y) \, dx \, dy.} \tag{4.14}$$

This formula works even for marginal expectations. For example,

$$\begin{aligned} E[Y] &= \int_y \int_x y \, f_{X,Y}(x,y) \, dx \, dy \\ &= \int_y y \left[\int_x f_{X,Y}(x,y) \, dx\right] dy \\ &= \int_y y \, f_Y(y) \, dy = \mu_Y. \end{aligned}$$

4.1.5.1 Covariance and Correlation

First and second order moments are key to our summary statistics. In the bivariate setting, there is a new second order moment, namely, the **covariance** of the r.v.s X and Y:

$$\boxed{\text{cov}(X, Y) = E[(X - \mu_X)(Y - \mu_Y)] \quad \textbf{covariance,}} \tag{4.15}$$

which intuitively measures the amount of variance shared by X and Y.

Example 1: If X and Y are independent, then the $\text{cov}(X, Y) = 0$, since $E[(X - \mu_X)(Y - \mu_Y)] = E[(X - \mu_X)] \cdot E[(Y - \mu_Y)] = 0 \cdot 0 = 0$. There is no variance shared by X and Y.

Example 2: If X and Y are linearly related, e.g., $Y = aX + b$, then $\mu_Y = a\mu_X + b$; hence, $Y - \mu_Y = (aX + b) - (a\mu_X + b) = a(X - \mu_X)$ and

$$\text{cov}(X, Y) = E[(X - \mu_X)(Y - \mu_X)] = E[(X - \mu_X) \cdot a(X - \mu_X)] = a\sigma_X^2.$$

The **correlation coefficient**, denoted by the Greek letter ρ, is the covariance of the standardized r.v.s X and Y:

$$\boxed{\text{cor}(X, Y) = E\left[\left(\frac{X - \mu_X}{\sigma_X}\right)\left(\frac{Y - \mu_Y}{\sigma_Y}\right)\right] \quad \textbf{correlation,}} \tag{4.16}$$

or in a more convenient form,

$$\boxed{\rho = \text{cor}(X, Y) = \frac{\text{cov}(X, Y)}{\sigma_X \sigma_Y} \quad \textbf{correlation coefficient.}} \tag{4.17}$$

Notice that ρ is a dimensionless quantity. If X and Y are already in standard form, then $\rho = E(XY)$.

Example 2 (continued): Since $Y - \mu_Y = a(X - \mu_X)$, then $\sigma_Y^2 = a^2\sigma_X^2$. Suppose $a \neq 0$, then

$$\rho = \frac{\text{cov}(X, Y)}{\sigma_X \sigma_Y} = \frac{a\sigma_X^2}{\sigma_X\sqrt{a^2 \cdot \sigma_X^2}} = \frac{a}{\sqrt{a^2}} = \text{sgn}(a) = \begin{cases} 1 & a > 0 \\ -1 & a < 0. \end{cases}$$

Thus linearly related random variables have correlation ± 1, depending upon the sign of the slope. We shall see that for any bivariate PDF, $-1 \leq \rho \leq 1$; hence, 100% of the variance is shared between X and Y in this case.

4.1.5.2 Regression Function

The regression function is a curve composed of all the conditional means:

$$\boxed{m(x) = E[Y|X = x] = \int_y y\, f_{Y|X=x}(y|x)\, dy \quad \textbf{regression function}} \tag{4.18}$$

Alternately, we can write $m(x) = \mu_{Y|X=x}$. For the example in Section 4.1.3,

$$m(x) = \frac{4}{5}x, \quad \text{a special case of } m(x) = \rho \cdot x,$$

where ρ is the correlation coefficient. The conditional variance is defined as

$$\boxed{\sigma_{Y|X=x}^2 = \int_y (y - \mu_{Y|X=x})^2\, f_{Y|X=x}(y|x)\, dy \quad \textbf{conditional variance,}} \tag{4.19}$$

which equals $1 - \rho^2$ for the standardized bivariate normal PDF. (Note: for the Pearson father–son height data, $\rho \approx 0.5$.)

4.1.6 Independence of *n* Random Variables

We generalize ideas of the previous section. We replace the (X, Y) notation with (X_1, X_2), which is easier to extend to *n* random variables.

$$F_{X_1, X_2, \ldots, X_n}(x_1, x_2, \ldots, x_n) = P(X_1 \leq x_1, X_2 < x_2, \ldots, X_n \leq x_n)$$
$$= P(X_1 \leq x_1) \cdot P(X_2 \leq x_2) \cdots P(X_n \leq x_n)$$
$$= F_{X_1}(x_1) \cdot F_{X_2}(x_2) \cdots F_{X_n}(x_n),$$

if the r.v.s $\{X_i\}$ are independent. The density function satisfies

$$f_{X_1, X_2, \ldots, X_n}(x_1, x_2, \ldots, x_n) = \frac{\partial^n F_{X_1, X_2, \ldots, X_n}(x_1, x_2, \ldots, x_n)}{\partial x_1 \, \partial x_2 \cdots \partial x_n}$$
$$= \frac{\partial^n [F_{X_1}(x_1) F_{X_2}(x_2) \cdots F_{X_n}(x_n)]}{\partial x_1 \, \partial x_2 \cdots \partial x_n}$$
$$= \prod_{j=1}^{n} \frac{\partial F_{X_j}(x_j)}{\partial x_j}, \text{ or}$$

$$\boxed{f_{X_1, X_2, \ldots, X_n}(x_1, x_2, \ldots, x_n) = \prod_{j=1}^{n} f_{X_j}(x_j) \quad \textbf{definition of independence of n r.v.s,}}$$

$$(4.20)$$

which we may take as the definition of independence of *n* random variables.

4.1.7 Bivariate Normal PDF

As in the univariate case, the bivariate normal is the exponential of a quadratic form. For simplicity, we assume the X and Y marginal random variables are in standard form, that is, their means are 0 and their variances are 1. Then the PDF is a function of only one parameter, ρ, and can be written as

$$\boxed{f_{X,Y}(x, y) = \frac{1}{2\pi \sqrt{1 - \rho^2}} \exp\left[-\frac{1}{2(1 - \rho^2)} (x^2 - 2\rho xy + y^2) \right].} \qquad (4.21)$$

We have seen this PDF in the example in Section 4.1.3. For a general choice of ρ, we may show that the marginal distributions of X and Y are both standard normal, and the conditional PDF of $Y|X = x$ is $N(\rho \cdot x, 1 - \rho^2)$; hence, the conditional variance is $1 - \rho^2$ for all x and

$$\boxed{f_{Y|X=x}(y|x) = \phi(y|\rho \cdot x, 1 - \rho^2) \quad \textbf{conditional PDF}} \qquad (4.22)$$

$$\boxed{m(x) = E[Y|X = x] = \rho \cdot x \quad \textbf{conditional mean,}} \qquad (4.23)$$

where $\phi(y|\mu_y, \sigma_Y^2)$ denotes the $N(\mu_Y, \sigma_Y^2)$ PDF. In that example, $\rho = 4/5$.

4.1.8 Correlation, Independence, and Confounding Variables

An important area of research in statistics is **causal inference**. For example, if there is a positive correlation between body weight and heart disease, can we conclude that being overweight causes heart disease? There are many, many examples of this type. It is not a statistical question whether such inferences are true. Usually other information and studies are required to confirm such a relationship.

Embarrassingly, R.A. Fisher was a heavy smoker and rejected correlations with such health outcomes as spurious. Indeed, there are correlations that are the result of a third, confounding variable. For example, when I was studying mosquito populations, there was a clear correlation (negative) with the phase of the moon. Did the moon really affect mosquito populations? No, mosquitoes were counted in "light traps," which were not as attractive to mosquitoes when there was a full moon.

Notice also that the correlation of X to Y is the same as Y to X. Could heart disease cause obesity? The question is not entirely frivolous, as heart disease will cause a decrease in physical activity, perhaps resulting in weight gain.

In any case, one must be cautious about drawing strong conclusions of causality from correlation. On the other hand, a lack of correlation certainly means there is no first order (linear) relationship.

Finally, the bivariate normal PDF has one very special and unique property; namely, if $\rho = 0$, then X and Y are both uncorrelated and independent as well. Equation (4.21) becomes

$$f_{X,Y}(x,y) = \frac{1}{2\pi}\exp\left[-\frac{1}{2}(x^2+y^2)\right]$$
$$= \frac{1}{\sqrt{2\pi}}\exp\left(-\frac{1}{2}x^2\right)\cdot\frac{1}{\sqrt{2\pi}}\exp\left(-\frac{1}{2}y^2\right)$$
$$= f_X(x)\cdot f_Y(y) ;$$

hence, by definition (4.13), X and Y are independent. Perhaps the causality question can be answered more definitively if the random variables are jointly Normal.

4.2 Change of Variables

Another avenue to interesting new distributions is by transformation of known distributions. Standardization is a trivial example. But we might be interested in the distribution of Z^2, the square of a standard normal random variable. In this section, we derive closed-form expressions for the density of such transformed random variables. We consider only the continuous case, as transformation of discrete random variables is just a matter of bookkeeping.

4.2.1 Examples: Two Uniform Transformations

As a preliminary step, we consider a direct method for finding the distribution of two transformations of a $X \sim \text{Unif}(0,1)$ random variable. Consider

$$Y_1 = X^2 \quad\text{and}\quad Y_2 = \sqrt{X}.$$

Intuitively, the transformations change outcomes from being "equally likely" to "more likely" to be closer to 0 or 1, respectively. In both cases, the domain and range are $(0, 1)$. Our approach is to find the CDF directly, that is,

$$F_{Y_1}(y_1) = P(Y_1 \le y_1) = P(X^2 \le y_1) = P(X \in (0, \sqrt{y_1}]) = \sqrt{y_1}$$

$$F_{Y_2}(y_2) = P(Y_2 \le y_2) = P(\sqrt{X} \le y_2) = P(X \in (0, y_2^2]) = y_2^2,$$

for y_j in the unit interval. Of course, $F_{Y_j}(y_j) = 0$ for $y_j < 0$ and $F_{Y_j}(y_j) = 1$ for $y_j > 1$. Because the domain is the positive interval $(0, 1)$, we choose to write $\sqrt{X^2}$ as X rather than $|X|$.

Finally, the PDFs are

$$f_{Y_1}(y_1) = \begin{cases} \dfrac{1}{2\sqrt{y_1}} & 0 \le y_1 \le 1 \\ 0 & y_1 \notin (0, 1) \end{cases} \quad \text{and} \quad f_{Y_2}(y_2) = \begin{cases} 2\,y_2 & 0 \le y_2 \le 1 \\ 0 & y_2 \notin (0, 1). \end{cases}$$

The first distribution is a Beta$(1/2, 1)$ PDF, and the second is an example of a triangular distribution. These are displayed in Figure 4.3.

4.2.2 One-Dimensional Transformations

In this section, we use the approach applied in Section 4.2.1 to obtain a general expression for the density, $g(y)$, of the transformed random variable, Y, from the original r.v. $X \sim f(x)$. (By using f and g for the PDFs, we avoid the necessity for subscripts X and Y.) We assume the transformation is a continuous and differentiable function on the X domain, which we take as a contiguous (single) interval, A. Furthermore, we assume that the function is monotone increasing on the interval A. This implies that the range is also an interval, call it B, and that the inverse transformation exists, at least between the domain and range. To keep the notation simple, we write the forward and inverse transformations as

$$\boxed{y = y(x),} \quad \text{which is 1–1, so that} \quad \boxed{x = x(y)} \quad \text{exists.}$$

Define a new contiguous interval $C \subset A$. Then the image under $y(x)$ of C is also a contiguous interval, call it D, which is a subset of B; that is, $D \subset B$; see Figure 4.4. The events $Y \in D$ and $X \in C$ occur simultaneously, and, therefore, have the same probability. We express this

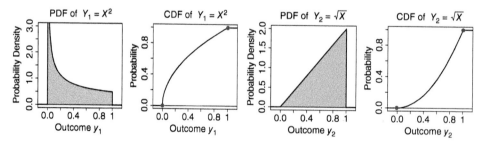

Figure 4.3 Transformations of a Unif $(0, 1)$ r.v.; see text.

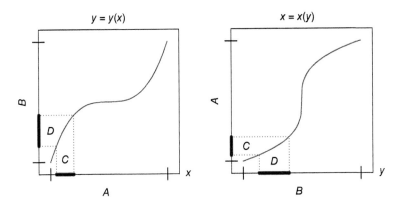

Figure 4.4 Sample transformations: $y = x^3$ and $x = \text{sgn}(y) \cdot |y|^{1/3}$. The range and domain of this transformation $A = B = (-1, 1)$.

probability as an integral and then perform the change of variables, $y = y(x)$,

$$\int_D g(y) \, dy = P(Y \in D) = P(X \in C) = \int_C f(x) \, dx$$

$$= \int_D f[x(y)] \, x(y)' \, dy \tag{4.24}$$

replacing x with the inverse transformation of x in terms of y, i.e., $x(y)$; replacing dx with $\frac{d\,x(y)}{dy}$, which is the Jacobian, and replacing the limits of integration C with D. Since the transformation is monotone increasing, both integration limits are in a positive order. Finally, since the derivation in Equation (4.24) holds for every interval C and D, the integrands of the first and last integrals must be equal; therefore, we have shown

$$g(y) = \begin{cases} f[x(y)] \, x(y)' & y \in B \\ 0 & y \notin B. \end{cases}$$

If the transformation happens to be monotone decreasing, a negative sign appears, but the order of integration is backwards; the general expression can be shown to be

$$\boxed{g(y) = \begin{cases} f[x(y)] \cdot |x(y)'| & y \in B \\ 0 & y \notin B. \end{cases}} \tag{4.25}$$

4.2.2.1 Example 1: Negative exponential PDF

Let $X \sim \text{Unif}(0, 1)$, and define $y(x) = -\log(x)$, which is a monotone decreasing function. Now $A = (0, 1)$, $B = (0, \infty)$, and $f(x) = 1$. Then the inverse transformation is $x(y) = \exp(-y)$ and the Jacobian $x(y)'$ is $-\exp(-y)$. Hence,

$$g(y) = f[e^{-y}] \cdot |-e^{-y}| = 1 \cdot e^{-y} = e^{-y}, \quad \text{for } y > 0,$$

which is a particular negative exponential density.

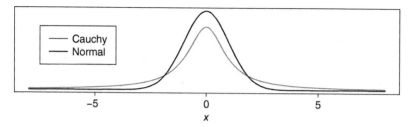

Figure 4.5 Standard Cauchy and normal PDFs.

4.2.2.2 Example 2: Cauchy PDF

Let $X \sim \text{Unif}(-\pi/2, \pi/2)$, and define $y(x) = \tan(x)$, which is a monotone increasing function. Now $A = (-\pi/2, \pi/2)$, $B = (-\infty, \infty)$, and $f(x) = 1/\pi$. Then the inverse transformation is $x(y) = \arctan(y)$ and the Jacobian $x(y)'$ is $1/(1 + y^2)$. Hence,

$$g(y) = f[\arctan y] \cdot \left| \frac{1}{1 + y^2} \right| = \frac{1}{\pi} \cdot \frac{1}{1 + y^2} \quad \text{so}$$

$$\boxed{g(y) = \frac{1}{\pi(1 + y^2)} \quad \text{for } y \in \mathbb{R}^1. \quad \text{Cauchy density}} \tag{4.26}$$

The Cauchy density in Figure 4.5 looks similar to the standard normal density, but with heavier tails (polynomial rather than exponential decrease). While the normal density is well behaved, the Cauchy is pathological, having no finite moments. It is one candidate for a "black swan" law; see Taleb (2007). Try the **R** command `rcauchy(100)` to see large values.

4.2.2.3 Example 3: Chi-squared PDF with one degree of freedom

Let $X \sim N(0, 1)$, and define $y(x) = x^2$; but $y(x)$ is not a monotone function from $A = (-\infty, \infty)$ to $B = (0, \infty)$. However, we can tweak the problem slightly to use our formula. Since the normal PDF is symmetric around 0 and $y(x) = |x|^2$, we now take $\tilde{A} = (0, \infty)$ and then take $f(x) = 2\phi(x)$ for $x \in \tilde{A}$. Now $y(x) = x^2$ is a 1–1 transformation from $\tilde{A}$ to B. Now the inverse transformation is $x(y) = +\sqrt{y}$ and the Jacobian $x(y)'$ is $1/2\sqrt{y}$. Hence,

$$g(y) = f[+\sqrt{y}] \cdot \left| \frac{1}{2\sqrt{y}} \right|$$

$$= 2 \cdot \frac{1}{\sqrt{2\pi}} \exp\left[-\frac{1}{2}(\sqrt{y})^2 \right] \cdot \frac{1}{2\sqrt{y}} \quad \text{or}$$

$$\boxed{g(y) = \frac{1}{\sqrt{2\pi}} y^{-1/2} e^{-y/2} \quad \text{for } y > 0, \quad \text{i.e., } Y \sim \chi^2(1),} \tag{4.27}$$

which is a particular chi-squared density, which is denoted by $Y \sim \chi^2(1)$.

4.2.3 Two-Dimensional Transformations

In this section, we extend the univariate transformation to a bivariate 1–1 transformation. We start with density $f(x_1, x_2)$ supported on a rectangle $A \subset \mathbb{R}^2$. The 1–1 transformation is defined by

$$\left.\begin{array}{l} y_1 = y_1(x_1, x_2) \\ y_2 = y_2(x_1, x_2) \end{array}\right\} \quad \text{inverse functions} \quad \left\{\begin{array}{l} x_1 = x_1(y_1, y_2) \\ x_2 = x_2(y_1, y_2) \end{array}\right.$$

and the range is a rectangle $B \subset \mathbb{R}^2$. The same change of variables derivation may be followed, with the result

$$\boxed{g(y_1, y_2) = f[x_1(y_1, y_2), x_2(y_1, y_2)] \cdot |J| \quad \text{for } (y_1, y_2) \in B,} \tag{4.28}$$

where $|J|$ denotes the absolute value of the Jacobian, J. J is defined as the determinant of the matrix of partial derivatives

$$J = \left\| \left(\begin{array}{cc} \dfrac{\partial x_1(y_1, y_2)}{\partial y_1} & \dfrac{\partial x_1(y_1, y_2)}{\partial y_2} \\ \dfrac{\partial x_2(y_1, y_2)}{\partial y_1} & \dfrac{\partial x_2(y_1, y_2)}{\partial y_2} \end{array} \right) \right\|. \tag{4.29}$$

Box–Müller transformation: Let $f(x_1, x_2) = 1$ on the unit square $A = (0, 1) \times (0, 1)$. This is the PDF of two independent Unif$(0, 1)$ random variables. Define the bivariate transformation

$$y_1 = y_1(x_1, x_2) = \sqrt{-2 \log(x_1)} \, \cos(2\pi x_2)$$

$$y_2 = y_2(x_1, x_2) = \sqrt{-2 \log(x_1)} \, \sin(2\pi x_2).$$

The range B is all of $\mathbb{R}^2$. The transformation is 1–1 except at the origin and infinity, which we may hope to ignore since they have probability 0. By summing the squares of y_1 and y_2, and then taking the ratio of y_2 and y_1, we can show the inverse functions are

$$x_1 = x_1(y_1, y_2) = \exp\left[-\frac{1}{2}(y_1^2 + y_2^2)\right]$$

$$x_2 = x_2(y_1, y_2) = \frac{1}{2\pi} \arctan\left(\frac{y_2}{y_1}\right).$$

Continuing, $g(y_1, y_2)$ is just the Jacobian, since $f(x_1(y_1, y_2), x_2(y_1, y_2)) = 1$:

$$J = \left\| \left(\begin{array}{cc} -y_1 \exp\left[-\frac{1}{2}(y_1^2 + y_2^2)\right] & -y_2 \exp\left[-\frac{1}{2}(y_1^2 + y_2^2)\right] \\ -y_2/[2\pi(y_1^2 + y_2^2)] & y_1/[2\pi(y_1^2 + y_2^2)] \end{array} \right) \right\|$$

$$= \frac{-y_1^2}{2\pi(y_1^2 + y_2^2)} \exp\left[-\frac{1}{2}(y_1^2 + y_2^2)\right] + \frac{-y_2^2}{2\pi(y_1^2 + y_2^2)} \exp\left[-\frac{1}{2}(y_1^2 + y_2^2)\right]$$

$$|J| = \frac{y_1^2 + y_2^2}{2\pi(y_1^2 + y_2^2)} \exp\left[-\frac{1}{2}(y_1^2 + y_2^2)\right]$$

$$= \frac{1}{\sqrt{2\pi}} \exp\left(-\frac{y_1^2}{2}\right) \cdot \frac{1}{\sqrt{2\pi}} \exp\left(-\frac{y_2^2}{2}\right) \quad \text{for } y_1 \in \mathbb{R}^1 \text{ and } y_2 \in \mathbb{R}^1,$$

which is the product of two univariate standard normal PDFs. Therefore, Y_1 and Y_2 are two independent $N(0, 1)$ random variables. How interesting and useful! See Box and Muller (1958).

4.3 Simulations

Running physical experiments can be time consuming and expensive. Computer experiments are increasingly powerful and can aid in the design of corresponding physical experiments. There is no physical model for predicting the future, so statistical modeling and computer experiments provide a unique capability. In this section, we describe the basics of running these so-called simulations.

4.3.1 Generating Uniform Pseudo-Random Numbers

Statistical models are driven by probabilistic behavior. How can such behavior be implemented on a computer? It was once believed that an external device, much like an external disk, could be designed to provide a stream of random digits. Such devices were built, and the CRC Standard Math Tables include a table of 1,000,000 random digits. These digits could be manipulated to produce random numbers with any desired properties. However, it soon became obvious that a million random digits was insufficient for almost any purpose.

4.3.1.1 Reproducibility

Fortunately, number theory came to the rescue and (deterministic) algorithms for generating **pseudo-random numbers** became available. One simple example that was published in the IMSL Library (Rogue Wave Software, currently Perforce Software) has the formula

$$x_{n+1} = 7^5 \cdot x_n \bmod (2^{31} - 1), \quad x_n = 1, 2, \ldots, 2^{31} - 2. \tag{4.30}$$

$$y_n = \frac{x_n}{2^{31} - 1} \approx \text{Unif}(0, 1). \tag{4.31}$$

This sequence is conveniently represented in 32-bit binary logic. It may be shown that if x_1 is any integer between 0 and $2^{31} - 1$, then the integer sequence defining x_{n+1} in Equation (4.30) hits every integer before it repeats. Normalizing by the modulus gives floating point numbers between 0 and 1 (excluding 0 and 1), as defined in Equation (4.31). However, the complete sequence of y_n is not uniformly spaced, as 32-bit floating point numbers have more decimal points near 0 than near 1. (This is the result of the fact that only 24 of the 32 bits are used to express the significant digits.)

The sequence (4.31) has been studied extensively, and no important non-random behavior has been discovered. Ironically, this sequence is deterministic, which is valuable when **debugging** computer codes. If every new run of a program used new random numbers, debugging would be iffy at best.

The sequence in Equation (4.30) can generate a billion random numbers. Today, even that is not considered adequate for all purposes. The random number generators used in **R** can produce virtually unlimited sequences before repeating. The details are left to the interested reader.

4.3.1.2 RANDU

Another similar sequence

$$x_{n+1} = 65539 \cdot x_n \quad \bmod \ 2^{31} \quad \text{and} \quad y_n = x_n / 2^{31} \tag{4.32}$$

was published as RANDU in the IBM Scientific Subroutine Package (SSP) in the 1960s as part of their Fortran offerings. The designers' goal was to make the prediction of x_{n+1} from x_n as difficult as possible. In doing so, they inadvertently made the prediction of x_{n+2} quite elementary. In fact, the three-dimensional points $\{(y_n, y_{n+1}, y_{n+2}), \ n = 1, 2, \dots \}$ can be shown to fall exactly on 15 equally spaced planes in the unit cube; see https://en.wikipedia.org/wiki/RANDU for a visualization.

4.3.2 Probability Integral Transformation

We have seen how to generate a sequence of pseudo-random numbers from the uniform distribution. We can use these to generate pairs of normal random variables using the Box–Müller transformation. For other continuous densities, $f(x)$, there is a remarkable result called the **probability integral transformation** (PIT), which requires only that the CDF, $F_X(x)$, has a nice and easily invertible formula.

Consider the PIT transformation

$$Y = F_X(X) \ ;$$

that is, we transform the random variable, X, by its own CDF. Thus Y takes on values between 0 and 1. What is the distribution of Y?

We use the direct method here; see Figure 4.6 as a visual aid.

$$P(Y \leq y) = P(F_X(X) \leq y) = P(X \leq F_X^{-1}(y))$$
$$= F_X(F_X^{-1}(y)) = y, \quad \text{for } 0 \leq y \leq 1.$$

We recognize the CDF of Y as the Unif$(0, 1)$ distribution. Thus, if $U \sim$ Unif$(0, 1)$, then $U = F_X(X)$; hence, $X = F_X^{-1}(U)$ can be used to transform uniform pseudo-random numbers to any continuous distribution desired. Note that F_X^{-1} denotes the inverse function, not the reciprocal operation. Thus, if $F_X^{-1}(u) = x$, then $F_X(x) = u$.

4.3.3 Event-driven Simulation

Many simulations, such as computer games, are driven by randomly occurring events that follow a Poisson process. Perhaps a character makes a random appearance at a rate λ.

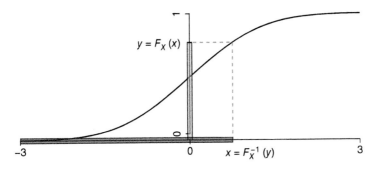

Figure 4.6 Generic PIT diagram. The red strip represents the event $Y \leq y$ while the blue strip represents the equivalent event $X \leq F_X^{-1}(y)$.

The classical way of coding up appearances of the character is to divide time into small increments, Δt. Then, by the Poisson axioms, the probability the character makes an appearance is approximately $p = \lambda \cdot \Delta t$. Each time the game clock advances one unit, we generate a new pseudo-random variable, u_n, and if the value is less than p, the character appears; otherwise, the character does not appear. Many games generate at least 30 frames per second, some as many as 100, and if the character appears only every 10 min on average, then p is very small and $u_n > p$ occurs most of the time.

An alternative is to use the fact that the distribution of time to the next Poisson event follows an exponential distribution. Thus we can use the probability integral transformation to generate one uniform number and transform it to the negative exponential scale. This is done by noting

$$F_T(t) = 1 - e^{-\lambda t} \qquad \text{the exponential CDF}$$

$$u = F_T(t) \qquad \text{the PIT equation}$$

$$u = 1 - e^{-\lambda t}$$

$$\boxed{t = -\frac{1}{\lambda}\log(1-u).}$$

This method could be much more efficient as a way of incrementing the game clock than a fixed time interval, Δt. Note that since the random variable $1 - U$ is also Unif(0, 1), we could use the formula $t = -\log(u)/\lambda$.

Problems

4.1 An alternative and informal derivation of Equation (4.5) may be illuminating. As we did in Figure 3.5, we write

$$P(X \approx x) = P\left(X \in \left(x - \frac{\Delta}{2}, x + \frac{\Delta}{2}\right)\right)$$

$$= P\left(X \in \left(x - \frac{\Delta}{2}, x + \frac{\Delta}{2}\right), Y \in (-\infty, \infty)\right)$$

$$= \int_{y=\infty}^{\infty} \left[\int_{s=x-\Delta/2}^{x+\Delta/2} f_{X,Y}(s,y)\, ds\right] dy$$

$$\approx \int_{y=\infty}^{\infty} [f_{X,Y}(x,y) \cdot \Delta]\, dy.$$

Since $P(X \approx x) \approx \Delta \cdot f_X(x)$, verify this sequence of equations and show that Equation (4.5) follows. Draw the relevant bivariate event.

4.2 Show that the sequence of pseudo-random variables $\{(x_n, x_{n+1}, x_{n+2}), n = 1, 2, \dots\}$ defined in Equation (4.32), satisfy the equation

$$x_{n+2} = 6\, x_{n+1} - 9\, x_n \quad \text{mod } 2^{31};$$

hence, the points (y_n, y_{n+1}, y_{n+2}), where $y_n = x_n/2^{31}$, fall on 15 planes in $\mathbb{R}^3$.

4.3 Use the bivariate change-of-variables approach to find the PDF of a bivariate normal, where the r.v.s are **not** in standard form. Conclude the correlation coefficient is unchanged. *Hint:* we know $f_{X,Y}(x,y)$ where X and Y are in standard form. Define two new r.v.s, $U = \mu_x + \sigma_x X$ and $V = \mu_y + \sigma_y Y$. Note that $U \sim N(\mu_X, \sigma_X^2)$ and $V \sim N(\mu_Y, \sigma_Y^2)$. Find $g_{U,V}(u,v)$, which is the general form of the bivariate normal PDF with all five parameters.

5

Approximations and Asymptotics

In Section 3.9.3, we saw how the Poisson distribution, Pois(m), converged to a normal distribution as its mean $m \to \infty$. These kinds of asymptotic results allow for useful approximations in statistical analysis and testing. Statistical analyses are based on a model of random data.

Statistics relies on the ability to replicate experiments many times independently in order to arrive at dependable estimates and decisions. This is captured in the idea of a random sample.

Random Sample: A collection of n random variables, $\{X_1, X_2, \ldots, X_n\}$, is defined to be a **random sample** if each random variable has the same distribution, that is,

$$X_i \sim F(x) \qquad \text{for } i = 1, 2, \ldots, n, \tag{5.1}$$

and if they are independent of each other. We say the $\{X_i\}$ are **independent and identically distributed (i.i.d.).**

When we form functions of a random sample, for example $S_n = \sum_{i=1}^{n} X_i$, we may wish to find approximations to the random variable S_n and understand its asymptotic limit and properties as $n \to \infty$. Probability theory gives us results in this vein and is the subject of this chapter.

5.1 Why Do We Like Random Samples?

Random samples have nice theoretical advantages. Modern data scientists often analyze huge datasets without sampling, so it is somewhat unclear what the generalization, if any, should be.

In a related vein, when designing experiments, Fisher pointed out that randomization could account for unknown influences by spreading them across the various treatment regimes. If two varieties of corn were to be compared by growing them side-by-side, it would be better to subdivide the acreage to compensate for local differences in topsoil quality, water availability, and shade. Knowing those influences and measuring them is preferred, but often we cannot know everything a priori.

Statistics: A Concise Mathematical Introduction for Students, Scientists, and Engineers, First Edition. David W. Scott.
© 2020 John Wiley & Sons Ltd. Published 2020 by John Wiley & Sons Ltd.

Recall that a random sample is a collection of n independent and identically distributed (i.i.d.) random variables. As we saw in Section 4.1.6, the joint PDF factors into the product of the marginals. Then a general expectation such as (letting $dx_1 \cdots dx_n$ be denoted by $d\mathbf{x}$)

$$E\,u(X_1, \ldots, X_n) = \int \cdots \int u(x_1, \ldots, x_n)\, f_{X_1, \ldots, X_n}(x_1, \ldots, x_n)\, d\mathbf{x}, \tag{5.2}$$

which invokes an n-dimensional integral, may be much easier to evaluate depending upon the exact form of the function $u(\cdot)$. We examine two special cases.

5.1.1 When $u(X)$ Takes a Product Form

The result in this section holds not only for random samples, but also for any collection of n independent random variables (not necessarily identically distributed). In either case, the joint PDF is the product of the n marginal PDFs. We prove the general result: if

$$u(X_1, \ldots, X_n) = \prod_{i=1}^{n} u_i(X_i)\,; \quad \text{then} \tag{5.3}$$

$$E\,u(X_1, \ldots, X_n) = \int_{\mathbb{R}^n} \left[\prod_{i=1}^{n} u_i(x_i) \right] f_{X_1, \ldots, X_n}(x_1, \ldots, x_n)\, dx_1 \cdots dx_n \tag{5.4}$$

$$= \int_{x_1} \cdots \int_{x_n} \left[\prod_{i=1}^{n} u_i(x_i)\, f_{X_i}(x_i)\, dx_i \right]$$

$$= \prod_{i=1}^{n} \int_{x_i} u_i(x_i)\, f_{X_i}(x_i)\, dx_i\,; \quad \text{hence,}$$

$$E\,u(X_1, \ldots, X_n) = \prod_{i=1}^{n} E\,u_i(X_i), \quad \text{when } u(\mathbf{X}) = \prod_{i=1}^{n} u_i(X_i). \tag{5.5}$$

Thus when $u(\mathbf{X})$ has the product form (5.3), we have converted a single n-dimensional integral in Equation (5.4) into n one-dimensional integrals in Equation (5.5). One way to remember this result is that *the expectation of the product is the product of the expectations.*

5.1.2 When $u(X)$ Takes a Summation Form

The result in this section holds not only for random samples, but for any collection of n random variables (not necessarily independent). Thus we again prove the general result. In the general form, we do not assume that the PDF $f(\mathbf{x})$ is the product of its marginal PDFs; but see Problem 1.

If $u(\mathbf{X})$ takes the form

$$u(X_1, \ldots, X_n) = \sum_{i=1}^{n} a_i\, u_i(X_i)\,; \quad \text{then} \tag{5.6}$$

$$E\,u(X_1, \ldots, X_n) = \int_{\mathbb{R}^n} \left[\sum_{i=1}^{n} a_i\, u_i(x_i) \right] f_{X_1, \ldots, X_n}(x_1, \ldots, x_n)\, d\mathbf{x}. \tag{5.7}$$

Recall that the marginal PDF is obtained by integrating out the other variables. Here for $\mathbf{x} \in \mathbb{R}^n$, that is summarized as

$$\int_{x_1} \cdots \int_{x_{i-1}} \int_{x_{i+1}} \cdots \int_{x_n} f_{\mathbf{X}}(\mathbf{x}) \, dx_1 \cdots dx_{i-1} \, dx_{i+1} \cdots dx_n = f_{X_i}(x_i),$$

which is the one-dimensional (marginal) PDF of X_i.

Continuing from Equation (5.7),

$$E\, u(\mathbf{X}) = \sum_{i=1}^{n} a_i \int_{\mathbb{R}^n} u_i(x_i) \, f_{\mathbf{X}}(\mathbf{x}) \, d\mathbf{x}$$

$$= \sum_{i=1}^{n} a_i \int_{x_i} u_i(x_i) \left[\int_{x_1} \cdots \int_{x_{i-1}} \int_{x_{i+1}} \cdots \int_{x_n} f_{\mathbf{X}}(\mathbf{x}) \right] d\mathbf{x}$$

$$= \sum_{i=1}^{n} a_i \int_{x_i} u_i(x_i) \, f_{X_i}(x_i) \, dx_i; \qquad \text{hence,}$$

$$\boxed{E\, u(\mathbf{X}) = \sum_{i=1}^{n} a_i \, E\, u_i(X_i), \quad \text{when } u(\mathbf{X}) = \sum_{i=1}^{n} a_i \, u_i(X_i).} \qquad (5.8)$$

Here, we have converted n n-dimensional integrals in Equation (5.7) into n one-dimensional integrals in Equation (5.8). One way to remember this result is that $\boxed{\textbf{\textit{the expectation of the}}}$ $\boxed{\textbf{\textit{sum is the sum of the expectations.}}}$

5.2 Useful Inequalities

To understand the properties of estimators and statistics as a function of the number of samples, n, a handful of inequalities help shed some light. We now derive some inequalities and their relevant applications.

5.2.1 Markov's Inequality

Consider a non-negative function, u, of a random variable, X, and any positive constant, $c > 0$; hence, the random variable $u(X) \geq 0$ satisfies

$$E[u(X)] = \int_{-\infty}^{\infty} u(x) \, f_X(x) \, dx$$

$$= \int_{-\infty}^{\infty} u(x) \underbrace{[I(u(x) \geq c) + I(u(x) < c)]}_{=1} f_X(x) \, dx$$

$$\geq \int_{-\infty}^{\infty} u(x) \, I(u(x) \geq c) \, f_X(x) \, dx \qquad \text{dropping the second term}$$

$$\geq \int_{-\infty}^{\infty} c \cdot I(u(x) \geq c) \, f_X(x) \, dx \qquad \text{since } u(x) \geq c \text{ here}$$

$$= c \cdot E[I(u(X) \geq c)] = c \cdot P[u(X) \geq c], \quad \text{using Eq (3.15); hence,}$$

$$\boxed{E[u(X)] \geq c \cdot P[u(X) \geq c] \qquad \textbf{Markov's inequality.}} \qquad (5.9)$$

Thus the expectation of a non-negative function of X is bounded by its right tail probability.

5.2.2 Chebyshev's Inequality

Markov's inequality leads to the more useful **Chebyshev's inequality**, which states there is a upper bound on the total probability a PDF has in the extreme tails, assuming that the r.v. has finite mean and variance. Let k be any positive constant and take $u(X) = (X - \mu_X)^2$ and $c = k^2 \sigma_X^2$ in Equation (5.9); then,

$$\sigma_X^2 = E[(X - \mu_X)^2] \geq \quad c \cdot P[(X - \mu_X)^2 > c]$$
$$= k^2 \sigma_X^2 \cdot P[(X - \mu_X)^2 > k^2 \sigma_X^2]$$
$$= k^2 \sigma_X^2 \cdot P(|X - \mu_X| > k\sigma_X).$$

Rearranging, we have shown

$$\boxed{P(|X - \mu_X| > k\sigma_X) \leq \frac{1}{k^2}, \qquad \textbf{Chebyshev's inequality.}} \qquad (5.10)$$

Any value of $k \leq 1$ is not interesting.

In Equation (3.51), we showed that $\mu_{\overline{X}} = \mu_X$ and $var\, \overline{X} = \sigma_{\overline{X}}^2 = \sigma_X^2/n$ for normal data. We will show in Section 6.2.1 that these formulae hold for any random sample. Hence, another version of Chebyshev's inequality is

$$\boxed{P(|\overline{X} - \mu_{\overline{X}}| > k\sigma_{\overline{X}}) = P\left(|\overline{X} - \mu_X| > k\frac{\sigma_X}{\sqrt{n}}\right) \leq \frac{1}{k^2} \,\,\textbf{Chebyshev's inequality.}} \quad (5.11)$$

Example: The probability a random variable takes on values more than 3 standard deviations away from its mean is always less than 1/9. For a standard normal distribution, this probability is 0.0027. Thus Chebyshev's inequality is usually not tight.

5.2.3 Jensen's Inequality[1]

For any r.v. X, if $g(x)$ is a convex function, then

$$\boxed{E[g(X)] \geq g(EX) \qquad \textbf{Jensen's inequality.}} \qquad (5.12)$$

Recall that the second derivative (if it exists) of a convex function $g(x)$ is positive (curves upward). In general, g is convex if for any two any points $x_1 < x_2$, the straight line connecting $g(x_1)$ and $g(x_2)$ lies above the curve $g(x)$ for $x_1 < x < x_2$. Furthermore, for each point x_0 in the domain of X, there is a line tangent to $g(x)$ at $x = x_0$ that lies below $g(x)$ for $x \neq x_0$. These two facts are illustrated in Figure 5.1.

To prove Jensen's inequality, define the tangent line at $x = EX$ to be $\ell(x)$; thus $\ell(x) = a + bx$ for some a and b. Since g is convex, we have $g(x) \geq \ell(x)$, and for $x = EX$, $g(EX) = \ell(EX)$; therefore,

$$g(x) \geq a + bx \qquad \text{since } g(x) \text{ is above the tangent line } \ell(x); \text{ hence,}$$
$$E[g(X)] \geq E(a + bX) \qquad \text{since expectation preserves ordering}$$
$$E[g(X)] \geq a + b\, EX = \ell(x = EX) = g(EX),$$

since the tangent line touches g at $x = EX$ by construction.

Example: One easy result follows if $g(x) = x^2$. Then $E(X^2) \geq (EX)^2 = \mu_X^2$. Of course, we know that $E(X^2) - \mu_X^2 = \sigma_X^2$ or $E(X^2) = \sigma_X^2 + \mu_X^2 \geq \mu_X^2$.

---

1 This section may be omitted at a first reading.

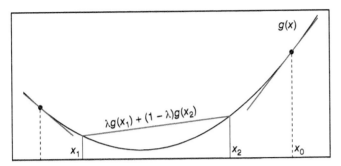

Figure 5.1 Example of a convex function $g(x)$ with two red tangent line segments touching the curve at the black points. A line segment connecting the curve at $x = x_1$ and $x = x_2$ is drawn in gray; see text.

5.2.4 Cauchy–Schwarz Inequality

The Cauchy–Schwarz inequality relates the second order moments of r.v.s X and Y:

$$|E(XY)| \leq E|XY| \leq \sqrt{EX^2}\sqrt{EY^2} \quad \textbf{Cauchy–Schwarz inequality.} \qquad (5.13)$$

To prove the first inequality, we must show that both $E(XY) \leq E|XY|$ and $-E(XY) \leq E|XY|$ are true. Clearly, $-|xy| \leq xy \leq |xy|$ for real numbers x and y. We first prove the following about certain functions g_1 and g_2:

If $g_1(xy) \leq g_2(xy) \ \forall \ x, y$, then $Eg_1(XY) \leq Eg_2(XY)$.

Since the joint density $f_{X,Y}(x, y)$ of the r.v.s is non-negative, then

$$g_1(xy)\, f_{X,Y}(x, y) \leq g_2(xy)\, f_{X,Y}(x, y); \quad \text{next integrating over } (x, y),$$
$$Eg_1(XY) \leq Eg_2(XY), \quad \text{proving the result.}$$

It follows that

$$xy \leq |xy| \quad \Longrightarrow \quad E(XY) \leq E|XY|$$
$$-xy \leq |xy| \quad \Longrightarrow \quad E(-XY) \leq E|XY| \quad \text{or}$$
$$- E(XY) \leq E|XY|$$

as was to be shown. (The same results hold if summing with discrete r.v.s.)

Now to prove the second inequality in Equation 5.13, define

$$U = \frac{|X|}{\sqrt{EX^2}} \quad \text{and} \quad V = \frac{|Y|}{\sqrt{EY^2}}.$$

Consider the polynomial $\frac{1}{2}U^2 + \frac{1}{2}V^2 - UV = \frac{1}{2}(U - V)^2 \geq 0$; hence,

$$\frac{1}{2}\frac{X^2}{EX^2} + \frac{1}{2}\frac{Y^2}{EY^2} \geq \frac{|X| \cdot |Y|}{\sqrt{EX^2}\sqrt{EY^2}} \ ; \quad \text{taking expectations implies}$$

$$\underbrace{\frac{1}{2}\frac{EX^2}{EX^2} + \frac{1}{2}\frac{EY^2}{EY^2}}_{=1} \geq \frac{E|XY|}{\sqrt{EX^2}\sqrt{EY^2}} \ ; \quad \text{rearranging, we get}$$

$$E|XY| \leq \sqrt{EX^2}\sqrt{EY^2}$$

as was to be shown, completing the proof.

Our main use for the Cauchy–Schwarz inequality is to show that the correlation coefficient ρ lies in the interval $[-1, 1]$; that is, $\rho^2 \leq 1$. The definition of ρ is

$$\rho = \frac{E[(X - \mu_x)(Y - \mu_y)]}{\sqrt{E(X - \mu_x)^2}\sqrt{E(Y - \mu_y)^2}} \quad \textbf{correlation coefficient.} \tag{5.14}$$

We may assume $\mu_x = \mu_y = 0$, w.l.o.g.; hence,

$$\rho^2 = \frac{[E(XY)]^2}{EX^2 \cdot EY^2} = \frac{|E(XY)|^2}{EX^2 \cdot EY^2}.$$

Starting with the two ends of Equation (5.13), we have

$$|E(XY)| \leq \sqrt{EX^2}\sqrt{EY^2}; \quad \text{squaring,}$$
$$|E(XY)|^2 \leq EX^2 \cdot EY^2, \quad \text{which implies}$$

$$\rho^2 = \frac{|E(XY)|^2}{EX^2 \cdot EY^2} \leq 1 \quad \text{Q.E.D.}$$

5.3 Sequences of Random Variables

Two random variables may be "close" to each other, i.e $X \approx Y$, in different ways. Sometimes two measurements taken at the same experiment are close to each other for every realization. Or it might be that only their distributions are similar, that is, $F_X \approx F_Y$. Intuitively, the first notion of closeness should also result in the second, but not vice versa. In this section, we examine how a sequence of random variables can approach another random variable. We start with several definitions of convergence.

Convergence of random variables: A sequence of random variables $X_1, X_2, X_3, \ldots$ is said to converge to a random variable X:

1. **In probability** (denoted by $X_n \xrightarrow{p} X$) if

$$\lim_{n \to \infty} P(|X_n - X| > \epsilon) = 0, \tag{5.15}$$

for every $\epsilon > 0$.

2. **In distribution** (denoted by $X_n \xrightarrow{D} X$) if

$$\lim_{n \to \infty} F_n(x) = F(x), \tag{5.16}$$

at all x where the CDF F is continuous.

3. **In mean square** (denoted by $X_n \xrightarrow{m.s.} X$) if

$$\lim_{n \to \infty} E(X_n - X)^2 = 0. \tag{5.17}$$

The sequence of random variables $X_1, X_2, X_3, \ldots$ does not generally denote a random sample as before, that is, they need not be *independent nor identically distributed*. To illustrate the difference, suppose $\overline{X}_n$ is the average of the first n r.v.s in a random sample; then we can generate a sequence as

$$\overline{X}_1, \overline{X}_2, \overline{X}_3, \ldots .$$

This sequence is clearly not identically nor independently distributed. The convergence of this sequence is the topic of the next section.

5.3.1 Weak Law of Large Numbers

The weak law of large numbers (WLLN) states that the sample mean $\overline{X}_n$ converges in probability to μ_X as $n \to \infty$. Given a random sample with $EX_i = \mu_X$ and var $X_i = \sigma_X^2 < \infty$ for all $i \geq 1$, define the sample mean $\overline{X}_n$ as the average of the first n random samples. Thus notationally, $X_1, X_2, \ldots$ is the random sample while $\overline{X}_1, \overline{X}_2, \ldots$ is the random sequence. As we will show in Section 6.2.1, the mean and variance of $\overline{X}_n$ are μ_X and σ_X^2/n, respectively, for any distribution (not just the normal PDF); therefore, choosing

$$\epsilon = k \frac{\sigma_X}{\sqrt{n}}, \quad \text{or} \quad \frac{1}{k} = \frac{\sigma_X}{\epsilon \sqrt{n}},$$

Chebyshev's inequality in Equation (5.11) becomes

$$P(|\overline{X}_n - \mu_X| \geq \epsilon) \leq \frac{\sigma_X^2}{\epsilon^2 n} \xrightarrow{n \to \infty} 0 \quad \text{for every } \epsilon > 0, \text{ which proves}$$

$$\boxed{\overline{X}_n \xrightarrow{p} \mu_X \qquad \text{WLLN} \,.} \tag{5.18}$$

5.3.2 Consistency of the Sample Variance

Under the same assumptions as in Section 5.3.1, define the sample variance of the first n samples as

$$\boxed{S_n^2 = \frac{1}{n-1} \sum_{i=1}^{n} (X_i - \overline{X}_n)^2.} \tag{5.19}$$

From the line before Equation 5.18, it follows that

$$P(|S_n^2 - \sigma_X^2| \geq \epsilon) \leq \frac{\text{var } S_n^2}{\epsilon^2} \xrightarrow{n \to \infty} 0 \quad \text{for every } \epsilon > 0,$$

if var $S_n^2 \to 0$ as $n \to \infty$. (This is discussed near Equation (6.9).) Note we replaced the variance of $\overline{X}_n$, which equals σ_X^2/n in the line before Equation (5.18), by the var S_n^2. Thus we have shown

$$\boxed{S_n^2 \xrightarrow{p} \sigma_X^2.} \tag{5.20}$$

5.3.3 Relationships Among the Modes of Convergence

The primary relationships among these three modes of convergence may be summarized as

$$\boxed{\text{(i) } X_n \xrightarrow{\text{m.s.}} X \;\;\Rightarrow\;\; X_n \xrightarrow{P} X.} \tag{5.21}$$

$$\boxed{\text{(ii) } X_n \xrightarrow{P} X \;\;\Rightarrow\;\; X_n \xrightarrow{D} X.} \tag{5.22}$$

5.3.3.1 Proof of Result (5.21)

Rewrite Markov's inequality in Equation (5.9) as

$$P[u(X_n - X) \geq \epsilon^2] \leq \frac{E\,[u(X_n - X)]}{\epsilon^2}.$$

Then, choosing u to be the non-negative squaring function, we have

$$P[(X_n - X)^2 \geq \epsilon^2] = P[|X_n - X| \geq \epsilon]$$

$$\leq \frac{E\,[(X_n - X)^2]}{\epsilon^2} \xrightarrow{n \to \infty} 0$$

by Markov's inequality and the definition of m.s. convergence in (5.17).

5.3.3.2 Proof of Result (5.22)[2]

We first show that for any r.v.s U and V, $a \in \mathbb{R}$, and $\epsilon > 0$, then

$$P(V \leq a) \leq P(U \leq a + \epsilon) + P(|V - U| > \epsilon). \tag{5.23}$$

This follows from

$$P(V \leq a) = P(V \leq a, U \leq a + \epsilon) + P(V \leq a, U > a + \epsilon) \quad \text{(partition)}$$

$$\leq P(\underbrace{U \leq a + \epsilon}_{\substack{\text{dropping } V \leq a \\ \text{constraint}}}) + P(V \leq a, U > a + \epsilon)$$

$$= P(U \leq a + \epsilon) + P(\underbrace{V - U \leq a - U}_{-U \text{ both sides}}, \underbrace{a - U < -\epsilon}_{\text{rearranging}})$$

$$\leq P(U \leq a + \epsilon) + P(V - U < -\epsilon), \quad \text{since } a - U < -\epsilon$$

$$\leq P(U \leq a + \epsilon) + \underbrace{P(V - U < -\epsilon) + P(V - U > \epsilon)}_{\text{extra positive term}},$$

which equals (5.23).

To prove (5.22), let x be any point where F (the CDF of X) is continuous. Then we make two applications of (5.23): first, let $V = X$, $U = X_n$, and $a = x - \epsilon$, giving

$$P(X \leq x - \epsilon) \leq P(X_n \leq \underbrace{(x - \epsilon) + \epsilon}_{\text{or just } x}) + P(|X - X_n| > \epsilon)\,;$$

second, let $V = X_n$, $U = X$, and $a = x$, giving

$$P(X_n \leq x) \leq P(X \leq x + \epsilon) + P(|X_n - X| > \epsilon).$$

---

2 This section may be omitted at a first reading.

Note that $P(X_n \le x)$ is common to both; stringing these together gives

$$P(X \le x - \epsilon) - P(|X - X_n| > \epsilon) \le P(X_n \le x) \le P(X \le x + \epsilon) + P(|X_n - X| > \epsilon).$$

Now let $n \to \infty$ carefully. Since (5.15) holds, we have

$$P(X \le x - \epsilon) - 0 \le \lim_{n\to\infty} P(X_n \le x) \le P(X \le x + \epsilon) + 0. \tag{5.24}$$

Now let $\epsilon \to 0$. Since $F(x) = P(X \le x)$ and is continuous at x, (5.24) becomes

$$P(X \le x) \le \lim_{n\to\infty} P(X_n \le x) \le P(X \le x) \quad \text{or}$$

$$\lim_{n\to\infty} P(X_n \le x) = P(X \le x).$$

This is equivalent to saying $F_n \to F$ everywhere that F is continuous. Therefore, we have shown $X_n \xrightarrow{D} X$.

5.4 Central Limit Theorem

The central limit theorem (CLT) states that standardized sums of i.i.d. random variables converge in distribution to a $N(0, 1)$ random variable. As in the special case of a Poisson random variable discussed in Section 3.9.3, the moment generating function is again at the heart of the proof. First, we examine the moment generating function for sums. We use the two results in Section 5.1 given in Equations (5.5) and (5.8) in our derivations.

5.4.1 Moment Generating Function for Sums

Let $S_n = \sum_{i=1}^{n} X_i$ be the sum of n i.i.d. random variables $X_1, \dots, X_n$. Then the moment generating function (MGF) of S_n is

$$M_{S_n}(t) = E[e^{tS_n}] = E[e^{t(X_1 + X_2 + \cdots + X_n)}]$$

$$= E[e^{tX_1} \cdot e^{tX_2} \cdots e^{tX_n}] = E[e^{tX_1}] \cdot E[e^{tX_2}] \cdots E[e^{tX_n}].$$

Since the samples are i.i.d., each has the same MGF, call it $M_X(t)$; hence,

$$\boxed{M_{S_n}(t) = [M_X(t)]^n} \qquad \text{MGF for the sum } S_n. \tag{5.25}$$

5.4.2 Standardizing the Sum S_n

Let μ_X and σ_X^2 be the moments of X_i. The mean of S_n is simply

$$ES_n = E(X_1 + X_2 + \cdots + X_n) = EX_1 + EX_2 + \cdots + EX_n \quad \text{or}$$

$$\boxed{\mu_{S_n} = ES_n = n\mu_X} \qquad \text{mean of the sum.} \tag{5.26}$$

The variance of S_n can be derived taking into account the independence of the $\{X_i\}$: var $S_n = E(S_n - ES_n)^2$, or

$$\text{var } S_n = E\left[\left(\sum_{i=1}^{n} X_i\right) - n\,\mu_X\right]^2 = E\left[\sum_{i=1}^{n}(X_i - \mu_X)\right]^2$$

$$= E\left[\sum_{i=1}^{n}(X_i - \mu_X)^2 + \sum\sum_{i\ne j}(X_i - \mu_X)(X_j - \mu_X)\right]$$

$$= \sum_{i=1}^{n} E[(X_i - \mu_X)^2] + \sum_{i \neq j} \sum E[(X_i - \mu_X)(X_j - \mu_X)]$$

$$= \sum_{i=1}^{n} \sigma_X^2 + \sum_{i \neq j} \sum E(X_i - \mu_X) \cdot E(X_j - \mu_X) = n\sigma_X^2 + 0, \quad \text{or}$$

$$\boxed{\sigma_{S_n}^2 = \text{var } S_n = n\sigma_X^2} \qquad \text{variance of the sum.} \tag{5.27}$$

Thus the standardized sum is given by sequence of random variables

$$\boxed{Y_n = \frac{S_n - n\mu_X}{\sigma_X \sqrt{n}} = \frac{n\overline{X}_n - n\mu_X}{\sigma_X \sqrt{n}}} \qquad \text{or} \tag{5.28}$$

$$\boxed{Y_n = \sqrt{n} \, \frac{\overline{X}_n - \mu_X}{\sigma_X}.} \tag{5.29}$$

We may write the CLT in several equivalent forms:

$$\boxed{\frac{1}{\sqrt{n}} \sum_{i=1}^{n} \frac{X_i - \mu_X}{\sigma_X} \xrightarrow{D} N(0,1) \quad \text{or} \quad \sqrt{n}\frac{\overline{X} - \mu_X}{\sigma_X} \xrightarrow{D} N(0,1) \quad \textbf{(CLT)}} \tag{5.30}$$

or moving σ_X to the right hand side,

$$\boxed{\frac{1}{\sqrt{n}} \sum_{i=1}^{n} (X_i - \mu_X) \xrightarrow{D} N(0, \sigma_X^2) \quad \text{or} \quad \sqrt{n}(\overline{X} - \mu_X) \xrightarrow{D} N(0, \sigma_X^2).} \tag{5.31}$$

5.4.3 Proof of Central Limit Theorem

Suppose the mean and variance of X_i are finite, that its MGF exists in a neighborhood of 0, and $Z_i = (X_i - \mu_X)/\sigma_X$ is the standard form of X_i; then

$$Y_n = \frac{S_n - n\mu_X}{\sigma_X \sqrt{n}} = \frac{\sum_{i=1}^{n}(X_i - \mu_X)}{\sigma_X \sqrt{n}} = \frac{1}{\sqrt{n}} \sum_{i=1}^{n} Z_i. \tag{5.32}$$

Using the standardized sum in the form given in Equation (5.32), we compute

$$M_{Y_n}(t) = E[e^{t \, Y_n}] = E\left[e^{\frac{t}{\sqrt{n}} \sum_{i=1}^{n} Z_i} \right]$$

$$= \prod_{i=1}^{n} E\left[e^{\frac{t}{\sqrt{n}} Z_i} \right] = \left[M_{Z_1}\left(\frac{t}{\sqrt{n}} \right) \right]^n. \tag{5.33}$$

Next we take a Taylor series of the MGF of Z_1 around $t = 0$, giving

$$M_{Z_1}\left(\frac{t}{\sqrt{n}}\right) = M_{Z_1}(0) + \frac{t}{\sqrt{n}}M'_{Z_1}(0) + \frac{1}{2}\left(\frac{t}{\sqrt{n}}\right)^2 M''_{Z_1}(0) + \cdots$$

$$= 1 + \frac{t}{\sqrt{n}} \cdot 0 + \frac{t^2}{2n} \cdot 1 + \cdots = \boxed{1 + \frac{t^2}{2n} + o(n^{-1}),} \qquad (5.34)$$

since $M'_{Z_1}(0) = EZ_1 = 0$ and $M''_{Z_1}(0) = EZ_1^2 = \sigma_{Z_1}^2 = 1$. Of course, $M_{Z_1}(0) = E(Z_1^0) = E[1] = 1$. Plugging Equation (5.34) into Equation (5.33),

$$M_{Y_n}(t) = \left[1 + \frac{t^2}{2n} + o(n^{-1})\right]^n \xrightarrow{n \to \infty} e^{t^2/2}, \qquad (5.35)$$

which is the MGF of a standard normal distribution; see Equation (3.46). This completes the proof of the CLT.

5.5 Delta Method and Variance-stabilizing Transformations

We begin by following an engineering approach to the delta method for a specific case. Suppose $X \sim \text{Pois}(m)$ and define $Y = \sqrt{X}$. We show

$$\boxed{Y = \sqrt{X} \approx N\left(\sqrt{m}, \tfrac{1}{4}\right),} \quad \text{a variance-stabilizing transformation.} \qquad (5.36)$$

Using the normal approximation to a Poisson PMF, we have $\text{Pois}(m) \approx N(m, m)$. Let Z denote a standard normal r.v.; then,

$$X \approx m + Z\sqrt{m} = m\left(1 + Z\frac{1}{\sqrt{m}}\right); \quad \text{hence,}$$

$$\sqrt{X} \approx \sqrt{m}\left(1 + Z\frac{1}{\sqrt{m}}\right)^{1/2} \approx \sqrt{m}\left(1 + Z\frac{1}{2\sqrt{m}} - Z^2\frac{1}{8m} + \cdots\right)$$

$$\approx \sqrt{m} + Z\frac{1}{2} + \cdots, \quad \text{where the next term is } \frac{-Z^2}{8\sqrt{m}} \approx 0 \text{ for large } m.$$

A $N(\mu_X, \sigma_X^2)$ random variable can be written exactly as $\mu_X + Z\sigma_X$; hence, we read off $\mu_X = \sqrt{m}$ and $\sigma_X = \frac{1}{2}$, which gives Equation (5.36). Since the square root transformation resulted in the variance being constant rather than some other function of m, it is called a **variance-stabilizing transformation**.

The same Taylor series approach can be used in general. We do not give the details of the so-called **delta method** here, but only the result. A proof can be found in more advanced textbooks.

Suppose Y_n converges in distribution to θ. Denote the transformation function by g and assume $g'(\theta) \neq 0$ exists. Specifically, we assume

$$\sqrt{n}(Y_n - \theta) \xrightarrow{D} N(0, \sigma^2) \qquad \text{for some } \sigma^2 \ ;$$

then,

$$\boxed{\sqrt{n}[g(Y_n) - g(\theta)] \xrightarrow{D} N(0, \sigma^2[g'(\theta)]^2) \qquad \textbf{delta method.}} \qquad (5.37)$$

Problems

5.1 Prove Equation (5.8) assuming $f(\mathbf{x}) = \prod_{i=1}^n f_{X_i}(x_i)$, i.e. $\{X_1, X_2, \dots, X_n\}$ is an i.i.d. sample. Note: this is a stronger assumption than required.

5.2 Show Chebyshev's inequality can also be expressed as

$$P(|X - \mu_X| > c) \leq \frac{\sigma_X^2}{c^2}.$$

5.3 Show the delta method given in Equation (5.37) for a Poisson r.v. with $g(x) = \sqrt{x}$ gives the same answer as in Equation (5.36).

6

Parameter Estimation

Each of the probability models we have studied so far has one or two parameters, θ or $\theta = (\theta_1, \theta_2)$, that must be specified. Given an experimental framework, a statistician collects a random sample of size n and attempts to estimate θ in an optimal fashion. This paradigm glosses over practical considerations, such as which form of the density to choose. The histogram, which can be described as a **nonparametric probability density estimator**, can provide guidance, as it can approximate almost any continuous density function. In the **R** function `hist`, use the `prob=T` option for proper scaling.

In this chapter, we will focus on **parametric probability density estimation**, assuming the statistician or subject matter expert has correctly selected the appropriate choice for the PDF, which we denote by $f_X(x|\theta)$. Therefore, the PDF for a realization of a random sample $\mathbf{X} = (X_1, X_2, \ldots, X_n)$ is the joint density

$$f_{X_1, X_2, \ldots, X_n}(x_1, x_2, \ldots, x_n|\theta) = \prod_{i=1}^{n} f_X(x_i|\theta) \ ; \tag{6.1}$$

since the random variables are i.i.d., we write $f_X(x_i|\theta)$ rather than $f_{X_i}(x_i|\theta)$. Sometimes we simply write the multivariate density as

$$f(x_1, x_2, \ldots, x_n|\theta) = \prod_{i=1}^{n} f(x_i|\theta), \tag{6.2}$$

where in Equation (6.2) we use the same symbol, f, to denote both the univariate and the multivariate densities. If there is a chance of confusion, we insert appropriate subscripts as in Equation (6.1). In many cases, where there is little chance of confusion, we may omit the subscripts. In that case, we rely on the argument(s) of the function to indicate its purpose.

In Figure 6.1, we display a random sample of size 50 from an unknown normal PDF, together with nine possible choices of the two parameters $\theta = (\mu, \sigma^2)$. It is easy to criticize some of these fits; for example, in the top left, the fit is too wide and shifted to the left. In other cases, the center is correct but the scale is wrong. Do you prefer the data displayed along the x-axis (as in the top row) or as a histogram? Which fit looks "best" to you?

Statistics: A Concise Mathematical Introduction for Students, Scientists, and Engineers, First Edition. David W. Scott.
© 2020 John Wiley & Sons Ltd. Published 2020 by John Wiley & Sons Ltd.

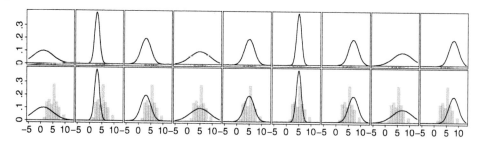

Figure 6.1 Nine examples of possible normal fits to a random sample of 50 points. In the first row, the data are displayed using the **R** function `rug(x)`. In the second row, probability histograms `hist(x,prob=T)` are displayed.

6.1 Desirable Properties of an Estimator

An estimator of a parameter θ is a formula using the random sample. As an example, consider the sample mean as an estimator, which we denote by $\hat{\theta}(X_1, X_2, \ldots, X_n) = \hat{\theta}_n = \overline{X}$. We know $E\overline{X} = \mu_X$. Therefore, if we are interested in estimating the mean, $\theta = \mu_X$, we call the estimator $\overline{X}$ **an unbiased estimator**. Such an estimator is accurate, at least on average.

We are also interested in finding estimators that are close to θ, on average. Thus if two estimators are both unbiased for θ, but one has smaller variance, we should prefer that one.

Finally, we would wish that as we collect more and more data,

$$\hat{\theta}_n \xrightarrow{n \to \infty} \theta$$

in probability, distribution, or mean square. In such cases, we say that $\hat{\theta}_n$ is **a consistent estimator of θ**.

Since the moments of a PDF will generally be a function of the parameter(s), we will focus on estimators of the first two moments. We have seen these before, but we reproduce them in the form of a random variable, and as a function of the random sample:

$$\overline{X} = \frac{1}{n} \sum_{i=1}^{n} X_i, \qquad \text{the **sample mean** r.v.} \tag{6.3}$$

$$S^2 = \frac{1}{n-1} \sum_{i=1}^{n} (X_i - \overline{X})^2, \qquad \text{the **sample variance** r.v.} \tag{6.4}$$

As before, lower case $\overline{x}$ and s^2 would be realizations of these random variables.

6.2 Moments of the Sample Mean and Variance

We begin by computing the *theoretical mean and theoretical variance* of the *sample mean*. In doing so, we learn how these random variables are related to the mean and variance of the sampling density. These calculations are completely general; however, we do assume that the moments exist (i.e., are finite) for the unknown density.

6.2.1 Theoretical Mean and Variance of the Sample Mean

We performed similar calculations in Section 5.4.2 when standardizing the random variable $S_n = \sum_{i=1}^{n} X_i = n \cdot \overline{X}$ or $\overline{X} = S_n/n$. (Note that the random variable S_n here is not related to S^2 or $\sqrt{S^2}$.) Recalling that $\mu_{S_n} = n \mu_X$ and $\operatorname{var} S_n = n \sigma_X^2$ from Equations (5.26) and (5.27), we have

$$E\overline{X} = \frac{1}{n} ES_n = \frac{1}{n}(n\mu_X) = \mu_X, \quad \text{i.e.} \quad \boxed{\mu_{\overline{X}} = \mu_X} \text{ and} \tag{6.5}$$

$$\operatorname{var} \overline{X} = E(\overline{X} - \mu_{\overline{X}})^2 = E(\overline{X} - \mu_X)^2 = E\left(\frac{S_n}{n} - \frac{\mu_{S_n}}{n}\right)^2$$

$$= \frac{1}{n^2} E(S_n - \mu_{S_n})^2 = \frac{1}{n^2} \cdot (n\sigma_X^2), \quad \text{i.e.} \quad \boxed{\sigma_{\overline{X}}^2 = \frac{\sigma_X^2}{n}.} \tag{6.6}$$

Thus we have the nice result that the sample mean is, on average, equal to the population mean. We also see that $\overline{X}$ must converge to μ_X in distribution by the WLLN, because its variance vanishes in the limit. But is this the best we can do? We will address this question later.

6.2.2 Theoretical Mean of the Sample Variance

Here we compute the average value of S^2 in Equation (6.4) in generality. Carefully organizing our work can greatly simplify the calculation. Since

$$(X_i - \mu_X) - \frac{1}{n} \sum_{j=1}^{n}(X_j - \mu_X) = (X_i - \mu_X) - \frac{1}{n}\left[\left(\sum_{j=1}^{n} X_j\right) - n\mu_X\right]$$

$$= X_i - \mu_X - \overline{X} + \mu_X = (X_i - \overline{X}),$$

we may assume without loss of generality that the random variables have been centered, that is, $\mu_X = 0$. With this assumption, we compute the expectation of S^2 from the inside out in Equation (6.4), starting with

$$X_i - \overline{X} = X_i - \frac{1}{n}\left[X_i + \sum_{j\neq i} X_j\right] = \frac{n-1}{n} X_i - \frac{1}{n} \sum_{j\neq i} X_j. \tag{6.7}$$

Squaring $(X_i - \overline{X})^2$ using Equation (6.7), we get three groups of terms:

$$\frac{(n-1)^2}{n^2} X_i^2 - \frac{2(n-1)}{n^2} X_i \sum_{j\neq i} X_j + \frac{1}{n^2}\left[\sum_{j\neq i} X_j^2 + \sum_{\substack{j\neq \ell \\ j\neq i,\ \ell\neq i}} \sum X_j X_\ell\right].$$

Since $\mu_X = 0$, $EX_i^2 = \sigma_X^2$ and $E[X_i X_j] = EX_i \cdot EX_j = 0$ for all $i \neq j$, the expectation of the previous expression is given by

$$\frac{(n-1)^2}{n^2} \sigma_X^2 - 0 + \frac{1}{n^2}[(n-1)\sigma_X^2 + 0] = \frac{(n-1)^2 + (n-1)}{n^2} \sigma_X^2 = \frac{n-1}{n} \sigma_X^2.$$

This is the expectation of $(X_i - \overline{X})^2$ for every i. Since S^2 has n terms with this expectation, and taking account of the factor $1/(n-1)$ in Equation (6.4), we have shown

$$ES^2 = \frac{1}{n-1} \cdot n \cdot \frac{n-1}{n} \sigma_X^2, \quad \text{i.e.} \quad \boxed{ES^2 = \mu_{S^2} = \sigma_X^2.} \tag{6.8}$$

Thus the formula for the sample variance (6.4) with denominator $n - 1$ is unbiased for $\sigma_{\bar{X}}^2$ for any sampling PDF with finite variance.

Notes: we shall often use the denominator n rather than $n - 1$ in the sample variance from time to time when we study maximum likelihood estimation. This estimator will be (slightly) biased. Also, if we are interested in an estimator for σ_X rather than σ_X^2, $\sqrt{S^2}$ will usually not be unbiased.

6.2.3 Theoretical Variance of the Sample Variance

We do not derive this expression, but note that it is given in exercise 10.13 of volume I of Stuart and Ord (1987). Assuming the required higher-order moments are finite,

$$\boxed{\text{var } S^2 = \frac{\kappa - \sigma_X^4}{n} + \frac{2}{n(n-1)}\sigma_X^4, \qquad \textbf{variance of the variance}} \qquad (6.9)$$

where κ (*kappa*) is the kurtosis of the sampling PDF. Notice that the second term is $O(n^{-2})$ and may be ignored. Equation (6.9) shows S^2 is consistent. For normal samples, $\kappa = 3\sigma_X^4$; hence, var $S^2 = 2\sigma_X^4/(n-1)$; see Problem 2.

6.3 Method of Moments (MoM)

The method of moments uses the intuitive idea from linear algebra that if our model density, $f(x|\theta)$, has p parameters, $\theta = (\theta_1, \theta_2, \ldots, \theta_p)$, then we may attempt to solve the p equations (perhaps linear, perhaps nonlinear) of the formulae for the first p sample moments, which will also be functions of the p parameters. Since the sample moments are consistent, we may reasonably expect that the parameter estimates will be consistent as well.

Example 6.1 Suppose the sampling density is the negative exponential $X \sim \exp(-x/\theta)/\theta$. Since $EX = \theta$, then $E\bar{X} = \theta$ as well; hence, $\hat{\theta} = \bar{X}$ is a MoM estimator. It is also the case that $\sigma_X^2 = \theta^2$, so that an alternative MoM estimator would be $\hat{\theta} = \sqrt{S^2}$. Since lower order moments can be more accurately estimated, we generally prefer using an estimator based upon the sample mean rather than the sample variance.

Example 6.2 Suppose the sampling density is the Gamma(α, β), for which $\mu = \alpha\beta$ and $\sigma^2 = \alpha\beta^2$. Then the MoM estimators for α and β may be found using the unbiased estimators $\bar{X}$ and S^2 for μ and σ^2 by solving the pair of nonlinear equations

$$\left.\begin{array}{l} \alpha\beta = \bar{X} \\ \alpha\beta^2 = S^2 \end{array}\right\} \implies \left\{\begin{array}{l} \hat{\alpha} = \dfrac{\bar{X}^2}{S^2} \\[2mm] \hat{\beta} = \dfrac{S^2}{\bar{X}}. \end{array}\right.$$

Notice we solved for $\hat{\beta}$ first by taking the ratio of the two equations.

Example 6.3 Our final example uses a simple-looking PDF,

$$f(x|\theta) = \theta x^{-2}, \qquad \text{for } x \geq \theta \text{ and some } \theta > 0. \tag{6.10}$$

It is easy to compute

$$\int_{\theta}^{\infty} \frac{\theta}{x^2} dx = -\frac{\theta}{x}\Big|_{x=\theta}^{\infty} = 1 \quad \text{but} \quad EX = \int_{\theta}^{\infty} \frac{x\theta}{x^2} dx = \theta \log x|_{x=\theta}^{\infty} = \infty \text{ ;}$$

that is, the mean does not exist for this PDF. So using $\overline{X}$ in the MoM will not work in this case. However, thinking outside the box, consider the first negative moment, EX^{-1}:

$$EX^{-1} = \int_{\theta}^{\infty} \frac{1}{x} \cdot \frac{\theta}{x^2} dx = -\frac{\theta}{2x^2}\Big|_{x=\theta}^{\infty} = \frac{1}{2\theta}.$$

It follows that

$$E\left[\frac{1}{X_1} + \frac{1}{X_2} + \cdots + \frac{1}{X_n}\right] = \frac{n}{2\theta}.$$

which suggests $\hat{\theta} = n/(2\sum_i X_i^{-1})$ as a possible MoM estimator. This density has surprisingly heavy tails; see Problem 3.

6.4 Sufficient Statistics and Data Compression

If we have a random sample from the PDF $f(x|\theta)$, the entire sample $\mathbf{X} = (X_1, X_2, \ldots, X_n)$ trivially contains **all the information** available for estimating the unknown parameter, θ, in the sampling density. In many cases, the joint density of the sample, called the likelihood function,

$$\boxed{L(\theta|\mathbf{X}) = f(X_1, .., X_n|\theta) = \prod_{i=1}^{n} f(X_i|\theta), \quad \textbf{likelihood function}} \tag{6.11}$$

simplifies in an interesting fashion. At $\mathbf{X} = \mathbf{x}$, the likelihood function is

$$\boxed{L(\theta|\mathbf{x}) = f(x_1, \ldots, x_n|\theta) = \prod_{i=1}^{n} f(x_i|\theta).} \tag{6.12}$$

$L(\theta|\mathbf{x})$ represents the likelihood function for a specific set of data, whereas $L(\theta|\mathbf{X})$ is a random variable representation of the likelihood of possible datasets.

Example 6.4 Suppose $X \sim \text{Binom}(1, p)$, which is a Bernoulli trial. Instead of writing the PMF as two cases, it can be written in the more compact form:

$$\boxed{p(x) = P(X = x|p) = \begin{cases} p & \text{if } x = 1 \\ 1 - p & \text{if } x = 0 \end{cases} = p^x(1-p)^{(1-x)}.} \tag{6.13}$$

We compute the likelihood function as

$$L(p|\mathbf{X}) = \prod_{i=1}^{n} p^{X_i}(1-p)^{1-X_i} = p^{\sum_{i=1}^{n} X_i}(1-p)^{n-\sum_{i=1}^{n} X_i}.$$

Thus the parameter, p, and the complete sample, $\mathbf{X}$, interact only through the data summary $\sum_{i=1}^{n} X_i$. This insight represents a significant data reduction or compression. Intuitively, it is also the case that the statistic $\overline{X} = \frac{1}{n} \sum X_i$ carries all the information in $\mathbf{X}$ about the parameter p.

Example 6.5 Suppose $X \sim N(\mu, \sigma^2)$. Then the likelihood function is

$$L(\mu, \sigma^2 | \mathbf{x}) = \prod_{i=1}^{n} \phi(x_i | \mu, \sigma^2) = \prod_{i=1}^{n} \frac{1}{\sqrt{2\pi\sigma^2}} \exp\left[-\frac{1}{2\sigma^2}(x_i - \mu)^2\right]$$

$$= (2\pi\sigma^2)^{-n/2} \exp\left[-\frac{1}{2\sigma^2} \sum_{i=1}^{n}(x_i - \mu)^2\right] \tag{6.14}$$

$$= (2\pi\sigma^2)^{-n/2} \exp\left[-\frac{1}{2\sigma^2}\left(\sum_{i=1}^{n} x_i^2 - 2\mu \sum_{i=1}^{n} x_i + n\mu^2\right)\right]$$

$$= (2\pi\sigma^2)^{-n/2} \exp\left(-\frac{n\mu^2}{2\sigma^2}\right) \exp\left[-\frac{1}{2\sigma^2} \sum_{i=1}^{n} x_i^2 + \frac{\mu}{\sigma^2} \sum_{i=1}^{n} x_i\right].$$

Thus the two parameters, μ and σ^2 interact with the data vector, $\mathbf{X}$, only through the summaries $\sum X_i$ and $\sum X_i^2$.

These examples lead us to define a sufficient statistic for a parameter.

Sufficient statistic(s): A statistic $T(\mathbf{X})$ is sufficient for θ if the conditional distribution $f(\mathbf{X} | T(\mathbf{X}) = t)$ does not depend on θ. It may be shown that $T(\mathbf{X})$ is a sufficient statistic if the likelihood function factors as

$$f(\mathbf{x}|\theta) = g(T(\mathbf{x})|\theta) \, h(\mathbf{x}). \tag{6.15}$$

This definition extends to the case where the number of parameters and sufficient statistics is greater than one, for example, the pair $(T_1(\mathbf{X}), T_2(\mathbf{X}))$ being sufficient for (θ_1, θ_2).

While it is still true that the entire data vector $\mathbf{X}$ is trivially sufficient for the parameters, sufficient statistics provide a useful lower-dimensional set of data summaries that also contain all the information about the parameter(s). We see from Examples 6.4 and 6.5 that $T_1(\mathbf{X}) = \sum_i X_i$ and $T_2(\mathbf{X}) = \sum_i X_i^2$ are sufficient statistics by the factorization theorem, which is proved in a more advanced course. Sufficient statistics exist in the discrete case as well, where we use the PMF in place of the PDF.

Example 6.4 (continued) We show the conditional distribution does not depend on p. The sufficient statistic is $T(\mathbf{X}) = \sum X_i$, which has a Binom(n, p) PMF. Now

$$p_{\mathbf{X}|T=t}\left(\mathbf{x} \Big| \sum x_i = t\right) = \frac{p_{(\mathbf{X},T=t)}\left(\mathbf{x}, \sum x_i = t\right)}{p_{T=t}\left(\sum x_i = t\right)}$$

$$= \frac{p^{\sum x_i}(1-p)^{n-\sum x_i}}{\binom{n}{t}p^t(1-p)^{n-t}} \quad \text{and} \quad \sum x_i = t$$

$$= \frac{p^t(1-p)^{n-t}}{\binom{n}{t}p^t(1-p)^{n-t}} = \frac{1}{\binom{n}{t}},$$

which does not have any further dependence (information) on the parameter p. Thus, $T(\mathbf{X}) = \sum X_i$ is the sufficient statistic.

6.5 Bayesian Parameter Estimation

The likelihood function has the data and parameter ordered as $f(\mathbf{x}|\theta)$. But we are interested in the reverse order. Perhaps Bayes theorem can guide us.

Let us use the symbol π to represent Bayesian PDFs. Formally, we write

$$\pi_{\Theta|\mathbf{X}}(\theta|\mathbf{x}) = \frac{f_{\mathbf{X}|\Theta}(\mathbf{x}|\theta)\,\pi_\Theta(\theta)}{f_{\mathbf{X}}(\mathbf{x})}. \tag{6.16}$$

For this to make any sense, θ cannot be a constant, but rather a random variable! A Bayesian statistician makes this assumption, while a frequentist statistician prefers to treat θ as a constant, although unknown. The focus of this text is on classical and frequentist methods. However, we provide one Bayesian example to hint at a follow-on course topic.

In Equation (6.16), the Bayesian statistician would call $\pi(\theta)$ the **prior distribution** of the unknown parameter. In specifying this PDF, the Bayesian can incorporate whatever information is available to them. If they have little specific information, a more diffuse choice for $\pi(\theta)$ would be appropriate.

On the left-hand side of Equation (6.16), the Bayesian statistician would call $\pi(\theta|\mathbf{x})$ the **posterior distribution** of the unknown parameter given the data. From this PDF, the Bayesian might pick a summary statistic and call it the **Bayesian estimator of θ**. The mean and mode of $\pi(\theta|\mathbf{x})$ are common choices. Two Bayesian statisticians analyzing the same dataset are likely to draw slightly different conclusions if they choose different prior distributions. For small n, this may be more of a practical issue.

Example with a Binomial Experiment: Suppose we have a random sample of size n from a $Binom(m, p)$. Note that we have changed the "n" in the binomial to "m" to avoid confusion. Our example analyzes data collected from 41 students who spun a nickel on its side and counted the number of heads in 25 repetitions; their data may be summarized in the sufficient statistic $\sum_{i=1}^{41} x_i = 588$. The Bayesian might believe that the unknown parameter, p, is 50%, the same as if the nickel had been flipped 25 times.

How do we choose the prior PDF $\pi(p)$? Picking the $Unif(0, 1)$ PDF would be satisfactory, but choosing the $Beta(\alpha, \beta)$ family not only includes that possibility when $\alpha = \beta = 1$, but also many more. (See the Appendix for the definition of a $Beta(\alpha, \beta)$ PDF.) Before we compute the posterior PDF per Equation (6.16), we note that the denominator, $f_{\mathbf{X}}(\mathbf{x})$, is a normalizing constant that does not involve θ. Thus we will compute the posterior initially ignoring factors that do not involve $\theta = p$.

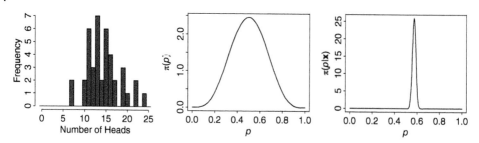

Figure 6.2 From left to right: a histogram of the 41 data points; the Beta(5, 5) prior PDF; and the posterior Beta(593,442) PDF.

First, we derive the expression for the likelihood function, noting

$$p_X(x|p) = \binom{m}{x} p^x q^{m-x}, \quad x = 0, 1, \dots, m,$$

recalling $q = 1 - p$. Hence,

$$p_X(\mathbf{x}|p) = \prod_{i=1}^{n} p_X(x_i|p)$$

$$\propto p^{\Sigma_i x_i} q^{nm-\Sigma_i x_i}.$$

Plugging into Equation (6.16) with the prior $\pi(p) \sim \text{Beta}(\alpha, \beta)$, we have

$$\pi(p|\mathbf{x}) \propto p^{\Sigma_i x_i} q^{nm-\Sigma_i x_i} \times p^{\alpha-1} q^{\beta-1}$$

$$= p^{\Sigma_i x_i + \alpha - 1} q^{nm-\Sigma_i x_i + \beta - 1}.$$

Fortunately, we recognize the posterior PDF as another Beta PDF, with parameters $\alpha' = \Sigma_i x_i + \alpha$ and $\beta' = nm - \Sigma_i x_i + \beta$. In Figure 6.2, we display the raw data and Bayesian results when $\alpha = \beta = 5$; hence, $\alpha' = 593$ and $\beta' = 442$. From the posterior PDF, we find the mode is $\hat{p} = 0.573$, and that the probability p is less than or equal to $\frac{1}{2}$ is only 1.3×10^{-6}. Flipping and spinning a nickel seem to exhibit different probabilistic behaviors. We note that the average number of heads was 14.34, or $\hat{p} = 0.574$; hence, the frequentist and Bayesian estimates are nearly identical. The Bayesian choosing a Unif(0, 1) *uninformative prior* would see $\hat{p} = 0.574$ and probability $p < \frac{1}{2}$ of 1.2×10^{-6}, only slightly different in this case.

6.6 Maximum Likelihood Parameter Estimation

The frequentist (non-Bayesian) statistician usually chooses the **maximum likelihood estimator (MLE)**, $\hat{\theta}$, which maximizes the likelihood function in Equation (6.11). Also, since the logarithm is a concave function, $\log(y) > \log(x) \iff y > x$, $\hat{\theta}$ also maximizes the log-likelihood function, which is defined as

$$\boxed{\ell(\theta|\mathbf{X}) \equiv \log L(\theta|\mathbf{X}) = \sum_{i=1}^{n} \log f(X_i|\theta) \quad \text{log-likelihood.}} \tag{6.17}$$

Summarizing, the MLE may be found in one of two ways:

$$\hat{\theta}_{\text{MLE}} = \arg\max_{\theta} L(\theta|\mathbf{X}) = \arg\max_{\theta} \prod_{i=1}^{n} f(X_i|\theta) \qquad \text{or} \qquad (6.18)$$

$$\hat{\theta}_{\text{MLE}} = \arg\max_{\theta} \ell(\theta|\mathbf{X}) = \arg\max_{\theta} \sum_{i=1}^{n} \log f(X_i|\theta). \qquad (6.19)$$

Often, the MLE can be derived by finding the point where the first derivative vanishes, and checking that the second derivative is negative there. If there is a constraint on the possible values of θ, the MLE may be at the boundary, which must be checked manually.

6.6.1 Relationship to Bayesian Parameter Estimation

Maximum likelihood treats the unknown parameter, θ, as a constant rather than as a random variable per Equation (6.16). However, if we take as the Bayesian point estimator the posterior mode of $\pi(\theta|x)$ while choosing the prior $\pi(\theta)$ as a constant (which is the uninformative prior), we see the MLE and Bayesian posterior mode are identical. If the support of the parameter θ is the real line, then the prior density cannot be Unif$(-\infty, \infty)$, since this density does not exist. However, in many cases, the point estimates for both methods are close, despite the great philosophical differences underlying each.

6.6.2 Poisson MLE Example

Since $\log f(x|m) = -m + x \log m - \log x!$, the log-likelihood is

$$\ell(m|\mathbf{X}) = \sum_{i=1}^{n} \log f(X_i|m)$$

$$= -nm + \log m \sum_{i=1}^{n} X_i - \sum_{i=1}^{n} \log(X_i!).$$

The MLE of $\theta = m$ is located where the first derivative vanishes:

$$0 = \frac{d}{dm}\ell(m|\mathbf{X}) = -n + \frac{1}{m}\sum_{i=1}^{n} X_i - 0 \quad \implies \quad \boxed{\hat{m} = \overline{X}.}$$

To verify that $\overline{X}$ is the MLE, we check that the second derivative is negative there. (We assume $\overline{X} > 0$, since it is the average of counts.)

$$\frac{d^2}{dm^2}\ell(m|\mathbf{X}) = 0 - \frac{1}{m^2}\sum_{i=1}^{n} X_i = -\frac{1}{m^2}n\overline{X}\Big|_{m=\overline{X}} = -\frac{n}{\overline{X}} < 0.$$

Note that the MLE and MoM estimators match in this case.

6.6.3 Normal MLE Example

From the normal likelihood given in Equation (6.14), writing $\theta_1 = \mu$ and $\theta_2 = \sigma^2$, we compute the log-likelihood as

$$\ell(\theta_1, \theta_2|\mathbf{X}) = -\frac{n}{2}\log 2\pi - \frac{n}{2}\log \theta_2 - \frac{1}{2\theta_2}\sum_{i=1}^{n}(X_i - \theta_1)^2.$$

The MLEs are located at a stationary point. First, we compute

$$\frac{\partial}{\partial \theta_1} \ell(\theta_1, \theta_2 | \mathbf{X}) = -\frac{1}{2\theta_2}(-2) \cdot \sum_{i=1}^{n}(X_i - \theta_1) = \frac{n}{\theta_2}(\overline{X} - \theta_1),$$

which vanishes when $\boxed{\hat{\theta}_1 = \overline{X}}$.

The MLE for the variance is located at a zero of

$$\frac{\partial}{\partial \theta_2} \ell(\theta_1, \theta_2 | \mathbf{X}) = -\frac{n}{2\theta_2} + \frac{1}{2\theta_2^2} \sum_{i=1}^{n}(X_i - \theta_1)^2,$$

which at $\hat{\theta}_1 = \overline{X}$ gives the MLE as

$$\boxed{\hat{\theta}_2 = \frac{1}{n}\sum_{i=1}^{n}(X_i - \overline{X})^2 = \frac{n-1}{n}S^2.} \tag{6.20}$$

Since S^2 is an unbiased estimator for σ^2 in general, we see the MLE here is biased slightly downwards.

To verify that $(\hat{\theta}_1, \hat{\theta}_2)$ are the MLEs, we must check that the matrix of second partial derivatives is negative definite there. Now

$$\frac{\partial^2}{\partial \theta_1^2} \ell(\theta_1, \theta_2 | \mathbf{X}) = -\frac{n}{2\theta_2}$$

$$\frac{\partial^2}{\partial \theta_2^2} \ell(\theta_1, \theta_2 | \mathbf{X}) = \frac{n}{2\theta_2^2} - \frac{1}{\theta_2^3} \sum_{i=1}^{n}(X_i - \theta_1)^2,$$

$$\frac{\partial^2}{\partial \theta_1 \partial \theta_2} \ell(\theta_1, \theta_2 | \mathbf{X}) = -\frac{n}{\theta_2^2}(\overline{X} - \theta_1).$$

Evaluating these at $\hat{\theta}$, we get the matrix

$$\begin{pmatrix} -n/(2\hat{\theta}_2) & 0 \\ 0 & -n/(2\hat{\theta}_2^2) \end{pmatrix},$$

which is clearly negative definite since both eigenvalues are negative.

6.6.4 Uniform MLE Example

Our final example is the uniform density, Unif$(0, \theta)$, where $\theta > 0$. We can write the density as

$$f(x|\theta) = \frac{1}{\theta} I(x \le \theta).$$

Hence, the likelihood function is

$$L(\theta|\mathbf{X}) = \prod_{i=1}^{n} f(X_i|\theta) = \begin{cases} \theta^{-n} & \text{if all } X_i \le \theta \\ 0 & \text{if any } X_i > \theta \end{cases}$$

$$= \theta^{-n} I(X_{(n)} \le \theta), \tag{6.21}$$

where $X_{(n)}$ is the largest **order statistic**. Now the unconstrained optimum for θ is 0, which gives an unbounded likelihood. However, the likelihood is actually 0 when θ is less than any of the data values. Hence, the constrained optimum value is the *smallest value* of θ that does not violate the constraint that $\theta \geq X_{(n)}$, namely,

$$\boxed{\hat{\theta} = X_{(n)}.}$$

This is an example where the calculus approach does not give the constrained optimum, which lies at the boundary. Examining Equation (6.21), we see that $X_{(n)}$ is the sufficient statistic by the factorization theorem.

6.7 Information Inequalities and the Cramér–Rao Lower Bound

Fisher (1922) was a strong advocate for the maximum likelihood paradigm. In this section, we outline how he was able to show that an ML estimator is asymptotically unbiased and asymptotically efficient. We show how Fisher defined the optimal information contained in each random sample and a bound on how small the variance of an estimator can be. The bound is called the **Cramér–Rao lower bound (CRLB)**. The technical details include assuming $f(x|\theta)$ and its derivatives are smooth, and that its support does not depend on θ.

6.7.1 Score Function

From Equation (6.19), the MLE $\hat{\theta}$ is a root of

$$\frac{\partial}{\partial \theta} \ell(\theta|\mathbf{X}) = \sum_{i=1}^{n} \frac{\partial}{\partial \theta} \log f(X_i|\theta). \tag{6.22}$$

This is the sum of i.i.d. random variables, so the central limit theorem applies to a properly scaled sum. Thus, we focus on a typical term in the sum, called the **score function**

$$\boxed{s(X;\theta) = \frac{\partial}{\partial \theta} \log f(X|\theta) = \frac{1}{f(X|\theta)} \frac{\partial f(X|\theta)}{\partial \theta} \qquad \textbf{score function.}} \tag{6.23}$$

The moments of the score function are

$$E[s(X;\theta)] = \int \left[\frac{\partial}{\partial \theta} \log f(x|\theta) \right] f(x|\theta) \, dx$$

$$= \int \left[\frac{1}{f(x|\theta)} \frac{\partial}{\partial \theta} f(x|\theta) \right] \cdot f(x|\theta) \, dx$$

$$= \int \frac{\partial}{\partial \theta} f(x|\theta) \, dx = \frac{\partial}{\partial \theta} \int f(x|\theta) \, dx = \frac{\partial}{\partial \theta} [1] = \boxed{0}; \tag{6.24}$$

and

$$\text{var}[s(X;\theta)] = E[s(X;\theta)^2] = \int \left[\frac{\partial}{\partial \theta} \log f(x|\theta) \right]^2 f(x|\theta) \, dx$$

$$= \int \left[\frac{\partial}{\partial \theta} \log f(x|\theta) \right] \left[\frac{\partial}{\partial \theta} \log f(x|\theta) \right] f(x|\theta) \, dx$$

$$= \int \left[\frac{\partial}{\partial \theta} \log f(x|\theta) \right] \frac{\partial}{\partial \theta} f(x|\theta) \, dx \quad \text{(see } E[s(X;\theta)] \text{ above)}$$

$$= -\int \left[\frac{\partial^2}{\partial \theta^2} \log f(x|\theta) \right] f(x|\theta) \, dx \quad \text{integrating by parts; or}$$

$$I_1(\theta) = -E\left[\frac{\partial^2}{\partial\theta^2}\log f(X|\theta)\right], \quad \text{Fisher information}. \tag{6.25}$$

Intuitively, the greater the curvature (more negative) of the score function for each sample, the greater the information each sample conveys.

For future reference, we note that by Equation (6.25),

$$\text{var}\left[\frac{\partial\ell(\theta|\mathbf{X})}{\partial\theta}\right] = \text{var}\left[\sum_{i=1}^{n} s(X_i|\theta)\right] = n \text{ var } s(X|\theta) = n\, I_1(\theta), \tag{6.26}$$

since the $s(X_i|\theta)$ are i.i.d. This is the Fisher information for the entire sample and is often denoted by $I_n(\theta)$.

Furthermore, we may note that Equation (6.25) also gives for $\mathbf{X}$

$$-E\left[\frac{\partial^2}{\partial\theta^2}\log f(\mathbf{X}|\theta)\right] = -E\left[\frac{\partial^2}{\partial\theta^2}\ell(\theta|\mathbf{X})\right] = -E\left[\sum_{i=1}^{n}\frac{\partial^2}{\partial\theta^2}\log f(X_i|\theta)\right]$$

$$= -\sum_{i=1}^{n} E\left[\frac{\partial^2}{\partial\theta^2}\log f(X_i|\theta)\right] = nI_1(\theta). \tag{6.27}$$

6.7.2 Asymptotics of the MLE

In this section, let $\hat{\theta}$ and θ denote the MLE and true parameter values, respectively. The log-likelihood derivative vanishes at the MLE; thus taking a Taylor series of $\ell(\hat{\theta}|\mathbf{X})$ around $\hat{\theta} = \theta$, we have approximately

$$0 = \frac{\partial\,\ell(\hat{\theta}|\mathbf{X})}{\partial\theta} = \frac{\partial\,\ell(\theta|\mathbf{X})}{\partial\theta} + (\hat{\theta} - \theta)\frac{\partial^2\,\ell(\theta|\mathbf{X})}{\partial\theta^2} + \cdots. \tag{6.28}$$

From Equation (6.22), the first term on right, $\partial\ell(\theta|\mathbf{X})/\partial\theta$, is the i.i.d. sum of the n score functions, $s(X_i; \theta)$, with mean 0 and variance $n \cdot I_1(\theta)$; see Equations (6.24) and (6.26) and Problem 1. By the CLT, $\partial\ell(\theta|\mathbf{X})/\partial\theta$ is approximately normal with mean 0 and variance $n\, I_1(\theta)$.

The last term on the right, $\partial^2\ell(\theta|\mathbf{X})/\partial\theta^2$, is the i.i.d. sum of the n r.v.s $\partial^2\log f(X_i|\theta)/\partial\theta^2$. By Equation (6.27) this has mean equal to $n\, I_1(\theta)$.

Next we represent a $N(\mu, \sigma^2)$ r.v. as $\mu + Z\sigma$, where Z is standard normal. Putting these together symbolically in Equation (6.28), we have

$$0 = [0 + Z\sqrt{n\, I_1(\theta)}] + (\hat{\theta} - \theta) \cdot [n\, I_1(\theta) + \cdots] + \cdots \quad \text{or}$$

$$n\, I_1(\theta) \cdot (\hat{\theta} - \theta) \approx 0 - Z \cdot \sqrt{n\, I_1(\theta)} \quad \text{or}$$

$$\sqrt{n}(\hat{\theta} - \theta) \approx 0 - Z \cdot \frac{1}{\sqrt{I_1(\theta)}} \quad \text{or}$$

$$\sqrt{n}(\hat{\theta} - \theta) \xrightarrow{D} N\left(0, \frac{1}{I_1(\theta)}\right), \tag{6.29}$$

since $-Z \sim N(0, 1)$ as well. This shows that the MLE is asymptotically unbiased, with asymptotic variance given by $1/(n\, I_1(\theta))$. In the next section, we show that the asymptotic variance of the MLE attains the theoretical lower bound.

6.7.3 Minimum Variance of Unbiased Estimators

Finally, we show that the minimum variance of any unbiased estimator of θ is exactly $(n\, I_1(\theta))^{-1}$. Since $\hat{\theta}(\mathbf{X})$ is unbiased, we have

$$\theta = \int \hat{\theta}(\mathbf{x})\, f(\mathbf{x}|\theta)\, d\mathbf{x}\,; \qquad \text{taking the } \partial/\partial\theta, \text{ we have}$$

$$1 = \int \hat{\theta}(\mathbf{x})\, \frac{\partial}{\partial\theta} f(\mathbf{x}|\theta)\, d\mathbf{x} = \int \hat{\theta}(\mathbf{x})\, \frac{\partial \log f(\mathbf{x}|\theta)}{\partial\theta} f(\mathbf{x}|\theta)\, d\mathbf{x} \quad \text{by (6.23)}$$

$$= \int (\hat{\theta}(\mathbf{x}) - \theta)\, \frac{\partial \log f(\mathbf{x}|\theta)}{\partial\theta} f(\mathbf{x}|\theta)\, d\mathbf{x} \quad \text{since } \theta \int \frac{\partial \log f}{\partial\theta} f = 0$$

$$\leq \sqrt{E[\hat{\theta}(\mathbf{X}) - \theta]^2} \cdot \sqrt{E\left[\frac{\partial \log f(\mathbf{X}|\theta)}{\partial\theta}\right]^2} \quad \text{by Cauchy–Schwarz,} \qquad (6.30)$$

where we have used the fact that the mean of the score function is zero to insert θ in the third line. Now

$$E\left[\frac{\partial \log f(\mathbf{X}|\theta)}{\partial\theta}\right]^2$$

is the variance of the i.i.d. sum in Equation (6.22), and we computed its variance in Equation (6.26) as $n\, I_1(\theta)$ as well. Squaring Equation (6.30), we have that

$$1 \leq \mathrm{var}(\hat{\theta}(\mathbf{X})) \cdot n\, I_1(\theta),$$

which re-arranged gives the inequality

$$\boxed{\mathrm{var}(\hat{\theta}(\mathbf{X})) \geq \frac{1}{n\, I_1(\theta)}, \quad \textbf{CRLB.}} \qquad (6.31)$$

The inequality in Equation (6.31) is called the $\boxed{\text{Cramér–Rao lower bound}}$.

Comparing Equations (6.29) and (6.31), we see that we have demonstrated Fisher's claim that the MLEs are asymptotically efficient. We remind the reader that there are cases that do not satisfy the smoothness or boundary assumptions. In those cases, the results may be quite different.

6.7.4 Examples

Example 6.6 For a Binom$(1, p)$ random sample, $\overline{X}$ is both the MLE and an unbiased estimator of p. The variance of $\overline{X}$ is $p(1-p)/n$. To compute the CRLB for this Bernoulli experiment, we note

$$\log p_X(X|p) = X \log p + (1-X)\log(1-p)$$

$$\frac{\partial^2}{\partial p^2} \log p_X(X|p) = -\frac{X}{p^2} - \frac{1-X}{(1-p)^2}; \quad \text{therefore,}$$

$$I_1(\theta) = -E\left[-\frac{X}{p^2} - \frac{1-X}{(1-p)^2}\right] = \frac{p}{p^2} + \frac{1-p}{(1-p)^2} = \frac{1}{p(1-p)}.$$

Hence, the MLE $\overline{X}$ achieves the CRLB, since $\mathrm{var}\,\overline{X} = \sigma_X^2/n = p(1-p)/n$.

Example 6.7 For a $N(\mu, \sigma^2)$ random sample, $\overline{X}$ is again both the MLE and an unbiased estimator of μ. The variance of $\overline{X}$ is σ^2/n. Computing the CRLB,

$$\log f(X|\mu, \sigma^2) = -\frac{1}{2}\log(2\pi\sigma^2) - \frac{1}{2\sigma^2}(X-\mu)^2 ; \quad \text{hence,}$$

$$\frac{\partial^2}{\partial\mu^2}\log f(X|\mu, \sigma^2) = -\frac{1}{\sigma^2} \quad \Longrightarrow \quad I_1(\theta) = \frac{1}{\sigma^2},$$

and the MLE again achieves the CRLB, since var $\overline{X} = \sigma^2/n$.

Example 6.8 Again for a $N(\mu, \sigma^2)$ random sample, we note that the MLE for σ^2 is not unbiased. However, we compute the CRLB for S^2 (which is unbiased) for its general interest.

$$\frac{\partial^2}{\partial(\sigma^2)^2}\log f(X|\mu, \sigma^2) = \frac{1}{2\sigma^4} - \frac{(X-\mu)^2}{\sigma^6} ; \quad \text{hence,} \tag{6.32}$$

$$I_1(\sigma^2) = -E\left[\frac{1}{2\sigma^4} - \frac{(X-\mu)^2}{\sigma^6}\right] = -\frac{1}{2\sigma^4} + \frac{\sigma^2}{\sigma^6} = \frac{1}{2\sigma^4}. \tag{6.33}$$

Now $\kappa = 3\sigma^4$. From Equation (6.9) var $S^2 = 2\sigma^4/(n-1)$, which exceeds the CRLB by the higher order term, $2\sigma^4/(n(n-1))$ in Equation (6.9).

Problems

6.1 Given n independent r.v.s X_i with means μ_i and variances σ_i^2, consider the random variable

$$Y = \sum_{i=1}^{n}(a_i X_i + b_i),$$

where a_i and b_i are known constants. Show

$$EY = \sum_{i=1}^{n}(a_i \mu_i + b_i) \tag{6.34}$$

$$\text{var } Y = \sum_{i=1}^{n} a_i^2\sigma_i^2. \tag{6.35}$$

Hint: define $Y_i = X_i - \mu_i$ and show $Y - \mu_Y = \sum a_i Y_i$.

6.2 For a normal random sample, show that var $S^2 = 2\sigma_X^4/(n-1)$ using the fact that $(n-1)S^2/\sigma_X^2 \sim \chi^2(n-1)$; see Equation (6.9) and following.

6.3 In the MoM Example 6.3 in Section 6.3, show that $F(x) = 1 - \theta/x$, so that we can generate random samples as $x = \theta/(1-u)$ by the probability integral transformation method, where u is a pseudo-independent *Unif* $(0,1)$ sample. Choose $\theta = 3$. For a sample of size 10^3, plot a histogram of the sample. Does the default **R** histogram show much structure? Does the alternative MoM estimator of θ seem to work? Verify that $\hat{\theta}$ is a MoM estimator.

6.4 Show that if the sampling density $f(x|\theta)$ has m sufficient statistics, then the MLE $\hat{\theta}$ must be a function of the data through the sufficient statistics alone.

7

Hypothesis Testing

In this chapter we explore the ways in which statisticians plan experiments, analyze data, and report findings. If every time we played a card game, we had to invent new rules, we would become quickly frustrated and not have the opportunity to improve our skills. By the same token, if every time we wished to perform a data analysis, we had to invent new algorithms and reporting language, we would frustrate other scientists trying to understand the findings.

Thus after a century of practical experience and trial-and-error, the scientific community has agreed upon paradigms or templates for analysis and reporting. As a result, the scientific community has become more efficient and effective in advancing the learning and decision-making process. Having standard rules available can be a great advantage. Certain disciplines, such as high-energy physics, have evolved even more restrictive rules, based upon their own experience mining petabytes of data for a handful of interesting events. They were noticing too many **false discoveries**. Their own modified rules have virtually eliminated that possibility.

We can see an example of standard statistical procedures and reporting by examining the author instructions from any medical journal. Take a few minutes to scan the author guidelines from the *Journal of the American Medical Association* (JAMA) https://jamanetwork.com/journals/jama/pages/instructions-for-authors. The issues are easily understood. We summarize a few salient points here (with key terms and ideas highlighted in bold for reference). Quoting from the heading **main outcome(s) and measures(s):**

> Indicate the primary study outcome measurement(s) as **planned before data collection began**. If the manuscript does not report the main planned outcomes of a study, this fact should be stated and the reason indicated. State clearly if the **hypothesis being tested** was formulated during or after data collection.

Here we see the expectation that the findings were not cherry picked, but were rather the result of careful planning to determine the effect of a drug or treatment regime. We also see how hypothesis tests should be reported.

About **reporting standards and data presentation**, JAMA writes

> When possible, present numerical results (eg, absolute numbers and/or rates) with appropriate indicators of **uncertainty**, such as **confidence intervals**. Use **means**

Statistics: A Concise Mathematical Introduction for Students, Scientists, and Engineers, First Edition. David W. Scott.
© 2020 John Wiley & Sons Ltd. Published 2020 by John Wiley & Sons Ltd.

and **standard deviations** (SDs) for normally distributed data and **medians** and **ranges** or **interquartile ranges** (IQRs) for data that are not normally distributed. Avoid solely reporting the results of statistical hypothesis testing, such as **P values**, which fail to convey important quantitative information. ... In general, there is no need to present the values of **test statistics** (eg, **F statistics** or χ^2 results) and **degrees of freedom** when reporting results.

... [T]here should be a description of how the potential for **type I error** due to **multiple comparisons** was handled, for example, by adjustment of the significance threshold. In the absence of some approach, these analyses should generally be described and interpreted as **exploratory**, as should all **post hoc analyses**.

We begin by describing the structure of common hypotheses tests. We also introduce the concept of a type I error.

7.1 Setting up a Hypothesis Test

The best illustration of the result of a well-designed experiment shows that a new technique/design/drug/intervention results in a change in the average response. The "shifted-normal model" can capture this well; see the left frame in Figure 7.1 for an example with both a shift in the mean and an increase in the standard deviation. This example is discussed in detail in Section 7.1.1.

The status quo is represented by the **null hypothesis**, and the hoped-for or anticipated improvement is represented by the **alternative hypothesis**:

$$H_0: \; X \sim f(x|\theta_0) \qquad \text{null hypothesis} \qquad\qquad (7.1)$$

$$H_1: \; X \sim f(x|\theta_1) \qquad \text{alternative hypothesis,} \qquad\qquad (7.2)$$

where f, θ_0, and θ_1 are all completely specified. H_0 and H_1 are referred to as **simple hypotheses**, since the two densities are completely specified by the given parameters, which are both single points (or vectors). Alternatively, we may consider a **composite hypothesis** for H_1, for example, $\theta_1 \in A$. If there is only one parameter, such a test might be that $\mu = \mu_0$ versus $\mu > \mu_0$ or $\mu \neq \mu_0$. These are referred to as **one-sided and two-sided alternative hypotheses**, respectively. In either case, the alternative sampling density is not only one choice of the PDF, but a family of possible choices.

As we read in the JAMA instructions, before n samples are collected, we must decide *for every* possible outcome sample vector $\mathbf{x} \in \mathbb{R}^n$ whether we are going to decide if H_0 or H_1 is the correct hypothesis. The set of points where we would decide that the hypothesis H_1 is correct is called the **critical region**, which we denote by the letter C.

7.1.1 Example of a Critical Region

As an example, suppose every house in a major hurricane sustained roof damage. The wind speed when the damage first occurred was measured, and the data followed a $N(70, 10^2)$ PDF using standard construction. An inventor built housing using a new truss tie-down fastener and claimed it performed better, with a $N(120, 15^2)$ distribution. However, a sample of

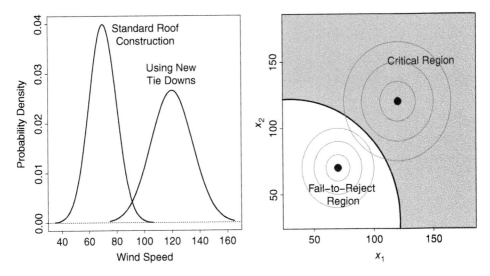

Figure 7.1   (Left) For two roof construction techniques, hypothetical PDFs $\phi(x|70, 10^2)$ and $\phi(x|120, 15^2)$ of the minimum wind speed incurred that resulted in roof damage during a hurricane. (Right) Illustration of a possible hypothesis-testing decision region for a small sample of $n = 2$ roofs. Contours of the two bivariate sampling PDFs are shown in green.

n houses using the new fastener may in fact perform no better than the current construction technique. Only if the sample data vector lands in the critical region do we reject the null hypothesis, deciding the weight-of-evidence favors the alternative hypothesis and supports the inventor's claim. An example using a very small sample of $n = 2$ is depicted in the right frame of Figure 7.1. Discussion of this example continues in Section 7.2.1.

7.1.2   Accuracy and Errors in Hypothesis Testing

A statistical test cannot be 100% accurate. We have two chances to be correct, and two ways of making an error. Given the asymmetric setup of the null and alternative hypotheses (status quo versus something new), it is generally a more serious error to reject the null hypothesis when it is true. For example, if a new $1000 pill is compared to an aspirin, then *incorrectly* deciding the new pill is better than an aspirin, which only costs a penny, is a very expensive mistake. This case is called a **type I error** or a **false positive outcome**.

The other possible error is to decide the status quo is correct when, in fact, the new method represents a real improvement. This is called a **type II error** or a **false negative outcome**. In this case, not enough data were collected to detect the improvement, or the improvement (while real) is not as large as hoped for. One of the planning activities in experimental design is to choose a sample size, n, sufficiently large to have a very good chance of detecting the (clinically relevant) improvement if it actually exists. The four cases are depicted in Table 7.1.

Often we have only one degree of freedom, so trading off type I and II errors is inevitable. While we could choose them to be equal, we shall see that these errors should not be treated symmetrically. Thus we always focus on the type I error, which we denote by α, choosing

Table 7.1 Truth table for hypothesis testing.

		Decision	
		H_0	H_1
Truth	H_0	✓	Type I error
	H_1	Type II error	✓

it to be a small number. Fisher advocated choosing $\alpha = 5\%$ or 1%, and then trying to set up the statistical analysis to make the type II error as small as possible. The probability of a type I error is called the **significance level of the test**. The complement of the type I error probability is called the **confidence level of the test**. The confidence level is generally 95% or 99% if we follow Fisher's guidance.

In the next sections, we describe a technique for finding the best critical region. Since the critical region induces a partition of the data space $\mathbb{R}^n$, we can estimate the probabilities of the four cases in Table 7.1 while planning.

7.2   Best Critical Region for Simple Hypotheses

The region where we decide to reject H_0 is called the **critical region**, which we denote by the letter C. Recall that the relative heights of a density function give the relative odds of the various outcomes. Intuitively, all points $\mathbf{x}$ on the boundary between the null and alternative hypotheses should have *exactly* the same relative odds. This intuition was proved as the Neyman–Pearson lemma (see Neyman and Pearson (1933)), which states that the form of the best critical region is defined by

$$C = \left\{ \mathbf{x} \in \mathbb{R}^n : \text{such that } \frac{f(\mathbf{x}|\theta_0)}{f(\mathbf{x}|\theta_1)} \le k \right\}, \tag{7.3}$$

where k is chosen to set the type I error at some desired level, usually 5% or 1%. Notice that the critical region includes those points $\mathbf{x}$ where the PDF under the null hypothesis has fallen sufficiently low compared to the alternative hypothesis. Equivalently, we may define C by

$$C = \{ \mathbf{x} \in \mathbb{R}^n : \text{such that } \log f(\mathbf{x}|\theta_0) - \log f(\mathbf{x}|\theta_1) \le \log(k) \}. \tag{7.4}$$

The ratio of the two PDFs in Equation (7.3) is called the **likelihood ratio**, while the difference in Equation (7.4) is the **log-likelihood ratio**. In both cases, the value of k determines the boundary between C and C^c.

7.2.1   Simple Example Continued

For the simple example shown in Figure 7.1, the most likely points under the null and alternative hypotheses are $(70, 70)$ and $(120,120)$, respectively, and are shown as blue dots in the right frame. Instead of specifying the type I error, k was chosen so that the boundary went through the point $(95, 95)$. A little algebra shows the boundary curve separating C and C^c satisfies

$$\frac{665}{36} + \frac{1}{6}(x_1 + x_2) - \frac{1}{360}(x_1^2 + x_2^2) = 0,$$

and that $\log(k) = -3.06676$. Since the two densities are well-separated, both errors are small:

$$\text{type I error} = \int_C \phi((x_1, x_2)|(70, 10^2)) \, dx_1 \, dx_2 = 0.0002$$

$$\text{type II error} = \int_{C^c} \phi((x_1, x_2)|(120, 15^2)) \, dx_1 \, dx_2 = 0.0089.$$

Thus in this example, a small dataset gives almost perfect decisions. However, while the data points (91,119) and (102,140) are both in the critical region, we might believe the claimed performance more in the latter case. Notice that we did not follow Fisher's guidance, although the type I error is smaller than the type II error.

7.2.2 Normal Shift Model with Common Variance

For this model, the simple hypotheses are

$$H_0: X \sim N(\mu_0, \sigma^2) \quad \text{versus} \quad H_1: X \sim N(\mu_1, \sigma^2), \tag{7.5}$$

where we assume σ^2 is known. The log-likelihood, which we denote by $\ell(\theta)$ or $\ell(\theta|\mathbf{x})$, for the jth hypothesis, $j = 0, 1$, is

$$\ell(\mu_j|\mathbf{x}, \sigma^2) = -\frac{n}{2} \log 2\pi\sigma^2 - \frac{1}{2\sigma^2} \sum_{i=1}^{n} (x_i - \mu_j)^2 \quad j = 0, 1. \tag{7.6}$$

Hence by Equation (7.4), the critical region is

$$\ell(\mu_0) - \ell(\mu_1) = -\frac{1}{2\sigma^2} \left[\sum_{i=1}^{n} (x_i - \mu_0)^2 - \sum_{i=1}^{n} (x_i - \mu_1)^2 \right]$$

$$= -\frac{1}{2\sigma^2} \sum_{i=1}^{n} [(x_i^2 - x_i^2) + 2(\mu_1 - \mu_0) x_i + (\mu_0^2 - \mu_1^2)]$$

$$= -\frac{n(\mu_1 - \mu_0)}{\sigma^2} \left[\bar{x} - \frac{\mu_0 + \mu_1}{2} \right] < \log k.$$

Thus the critical region takes the r.v. form (noting the positive or negative sign of the factor $-(\mu_1 - \mu_0)$ at the beginning of the right hand side):

$$\left. \begin{array}{l} \overline{X} > k' \quad \text{if } \mu_1 > \mu_0 \\ \overline{X} < k' \quad \text{if } \mu_1 < \mu_0 \end{array} \right\} \quad \text{where} \quad \overline{X} \sim N\left(\mu_0, \frac{\sigma^2}{n}\right). \tag{7.7}$$

Example: Let $\mu_0 = 0$, $\mu_1 = 10$, $\sigma = 24$, with $n = 9$ or $n = 64$. We shall pick k' so that the type I error is 5%. Thus we wish

$$\alpha = P(\overline{X} > k' \mid \mu = \mu_0 = 0) = P\left(\frac{\overline{X} - 0}{24/\sqrt{n}} > \frac{k' - 0}{24/\sqrt{n}}\right)$$

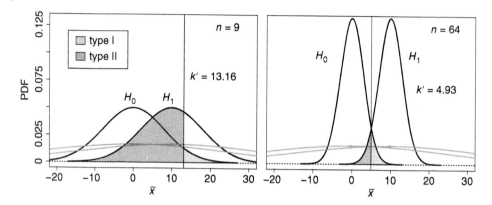

Figure 7.2 Critical regions based upon $\overline{X} > k'$ for testing two shifted normal PDFs with $\mu_0 = 0$, $\mu_1 = 10$, and common $\sigma = 24$. (Left) $n = 9$; (right) $n = 64$. The type I and II errors are shown in red and blue, respectively. The underlying sampling densities and means are shown in green; the densities for $\overline{X}$ are shown in black.

$$\alpha = P\left(Z > \frac{k'\sqrt{n}}{24}\right) \implies \frac{k'\sqrt{n}}{24} = \Phi^{-1}(1-\alpha)$$

so $k' = 13.159$ and 4.935, respectively, for $n = 9$ and 64; see Figure 7.2.

Let β denote the probability of correctly rejecting H_0 when H_0 is false; see Section 7.4.3. Then given k', the type II error is its complement, or $1 - \beta$, which is computed as

$$1 - \beta = P(\overline{X} < k' \mid \mu = \mu_1 = 10) = P\left(\frac{\overline{X} - 10}{24/\sqrt{n}} < \frac{k' - 10}{24/\sqrt{n}}\right)$$

$$1 - \beta = P\left(Z < \frac{(k'-10)\sqrt{n}}{24}\right) = \Phi\left(\frac{(k'-10)\sqrt{n}}{24}\right), \tag{7.8}$$

or 65.49% and 4.57%, respectively. These are computed as $P(\overline{X} \in C^c \mid \mu = 10)$ or $P(\overline{X} < k' \mid \mu = 10)$. Again, the type I errors, $P(\overline{X} > k')$, are both 5% by choice. These are also depicted in Figure 7.2. Observe that as the sample size increases, the two shifted normal PDFs for $\overline{X}$ overlap less.

7.3 Best Critical Region for a Composite Alternative Hypothesis

With a composite hypothesis or a simple hypothesis that does not specify all parameters, the critical regions defined in Equations (7.3) and (7.4) are modified so that any unknown parameters are replaced by their maximum likelihood values. If any parameters are specified in the null or alternative hypotheses, then the MLEs should be constrained rather than unconstrained.

7.3.1   Negative Exponential Composite Hypothesis Test

The negative exponential density, $T \sim f(t|\beta) = \beta^{-1}e^{-t/\beta}$, is a special case of the two-parameter Gamma PDF, namely Gamma$(1, \beta)$, with $\beta > 0$. We wish to perform a test with a composite alternative

$$H_0: \beta = \beta_0 \quad \text{versus} \quad H_1: \beta \neq \beta_0. \tag{7.9}$$

Note $\log f(t|\beta) = -\log \beta - t/\beta$. Writing $s_n = \sum_{i=1}^{n} t_i$, the log-likelihood is

$$\ell(\beta|\mathbf{t}) = -n \log \beta - \frac{1}{\beta}\sum_{i=1}^{n} t_i = -n \log \beta - \frac{s_n}{\beta},$$

for which the MLE is clearly $\hat{\beta} = s_n/n = \bar{t}$. Since the alternative hypothesis does not specify the density, we replace the unknown parameter β_1 with its MLE, namely, $\bar{t} = s_n/n$. Hence, the log-likelihood ratio is given by

$$\ell(\beta_0) - \ell(\beta_1 = s_n/n) = \left[-n \log \beta_0 - \frac{s_n}{\beta_0} \right] - \left[-n \log(s_n/n) - \frac{s_n}{s_n/n} \right]$$

$$= -n \log \beta_0 - \frac{s_n}{\beta_0} + n \log s_n - n \log n + n.$$

We give some examples in the next sections.

7.3.1.1   Example

Suppose $\beta_0 = 1$ and $n = 8$. In the left frame of Figure 7.3, we plot the log likelihood ratio as a function of s_8. The critical region is clearly of the form $a < s_8 < b$. Finding a and b requires quite a bit of work in the form of nonlinear root calculations.

In general, the test is based upon the random variable, $s_n = \sum_{i=1}^{n} T_i$. Since each sample, T_i, is Gamma$(1, \beta_0)$, a quick look at the form of the MGF of the Gamma PDF shows its product results in S_n following the Gamma(n, β_0) exactly. In the right frame of Figure 7.3, we show this PDF, together with the 5% tail region shaded in red.

7.3.1.2   Alternative Critical Regions

The critical region displayed in Figure 7.3 is not the only choice we might contemplate. Any interval satisfying the significance level of 5% can be considered. If we view the result of the log likelihood ratio as revealing that the random variable S_n is the key quantity, then there are two more choices of (a, b) that we might prefer.

The first choice is the equal-tail-area option. That is, we choose a and b to be the 2.5th and 97.5th percentiles, respectively, of the random variable S_n. For our example, $(a, b) = (3.45, 14.42)$; see Figure 7.4(Left).

The second choice is to choose (a, b) so that the width of the interval, $b - a$, is as small as possible. In Problem 2, we learn that the heights $f(a)$ and $f(b)$ must be equal in this case, not unequal as in Figure 7.3. Numerical search finds that $(a, b) = (2.97, 13.63)$; see Figure 7.4(Right).

The widths of the three 95% significance test intervals are 11.37, 10.97, and 10.66. The first two are only 6.64% and 2.92% wider. Perhaps the equal-tail-area strategy is easiest and does not pay a great penalty. See also Problem 3 for another justification for choosing equal tail areas.

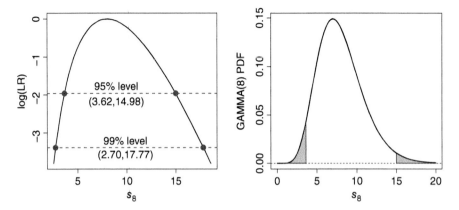

Figure 7.3 (Left) The log-likelihood ratio for a sample of $n = 8$ negative exponential r.v.s with $\beta_0 = 1$. The levels corresponding to 5% and 1% type I errors are shown. (Right) The Gamma$(8, \beta = 1)$ PDF of S_8, together with the 95% probability interval $(3.62, 14.98)$. The shaded tail areas have mass 3.176% and 1.824%, respectively, totaling 5%.

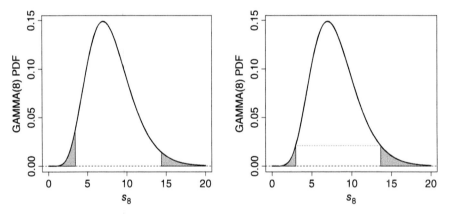

Figure 7.4 Alternative 95% confidence level tests for our example. (Left) $(3.45, 14.42)$ has equal tail probabilities of 2.5%; (right) $(2.97, 13.63)$ is the narrowest interval. The tail areas are 1.824% and 3.176%, respectively.

7.3.1.3 Mount St. Helens Example

We give an example of the negative exponential composite test in Equation (7.9) using data collected over the ten days before the 19 March 1982, eruption of Mount St. Helens. The time stamp (in days since 1 January 1982) and epicenters of 247 earthquakes were recorded; see Figure 7.5 for a spatial view of these data points. In Figure 7.6, histograms of the times between eruptions are displayed, together with MLE fits of a negative exponential PDF. The average time interval was $\beta = 0.0581$ days for the first 147 eruptions. We wish to test if this parameter is satisfactory for the last 100 eruptions. The data are the 99 time intervals, denoted by $\{T_i\}$. Following Section 7.3.1.1, the test is based upon the statistic

$$S_{99} = \sum_{i=1}^{99} T_i \sim \text{Gamma}(99, 0.0581).$$

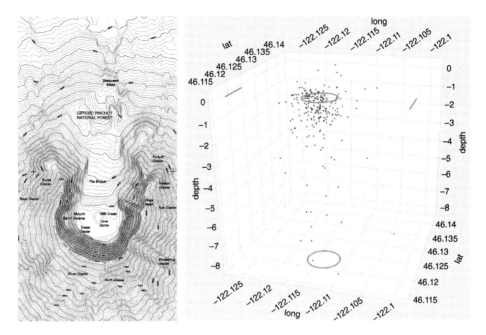

Figure 7.5 (Left) Topo map; (right) earthquake epicenters.

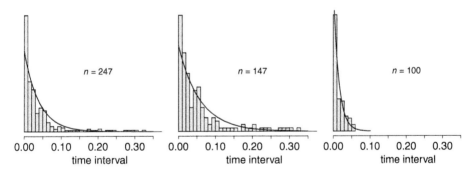

Figure 7.6 Histograms of times between eruptions (in days) for all 247 eruptions (left frame), the first 147 eruptions (middle frame), and last 100 eruptions (right frame). The blue line depicts a negative exponential fit.

Using the equal-tail-area convention, the critical region at the 5% level is the complement of interval

$$(\text{Gamma}_{0.025}(99, 0.0581), \text{Gamma}_{0.975}(99, 0.0581)) = (4.67, 6.94).$$

For these data, $s_{99} = 1.43$, which is in the critical region. Hence, we reject H_0 at the 5% level. Examining Figure 7.6, we see the frequency of earthquakes dramatically increased before the eruption.

7.3.2 Normal Shift Model with Common But Unknown Variance: The *T*-test

In Equation (7.7), we saw that the log-likelihood ratio test showed the standard normal random variable

$$Z = \frac{\overline{X} - \mu_0}{\sigma/\sqrt{n}}$$

was used to determine the critical region, *when the variance σ^2 was assumed to be known.* That is an unusual situation (except, perhaps, in textbooks).

On the other hand, if σ^2 is in fact unknown, we might expect the other sufficient statistic, S^2 to enter into the log likelihood ratio test for the composite alternative in

$$H_0: \mu = \mu_0 \quad \text{versus} \quad H_1: \mu \neq \mu_0, \quad \text{assuming } X \sim N(\mu, \sigma^2), \tag{7.10}$$

where σ^2 is unknown. We derive the relevant r.v., and show it is

$$T_{n-1} = \frac{\overline{X} - \mu_0}{\sqrt{S^2/n}} \tag{7.11}$$

that performs that role in the LR test.

Aside: this is the famous *T*-test, or Student's *T*-test, published by a statistician named William Gosset; see "Student" (1908). Gosset, who was employed by Guinness Brewery, developed this test, which is more appropriate for small sample sizes than the *Z*-test. Guinness allowed its employee to publish the result in Pearson's Biometrika in 1908, but only under the pseudonym "Student" so competitors would not immediately see its usefulness in improving the quality of their beer (stout) production.

We begin by understanding the *T* random variable, its distribution and form. Then we derive the log likelihood ratio critical region.

7.3.3 The Random Variable T_{n-1}

The random variable T_{n-1} in Equation (7.11) involves both $\overline{X}$ and S^2. For normal data, we know the distribution of $\overline{X}$ is $N(\mu, \sigma^2/n)$ exactly. In this section, we find the exact sampling distribution of the sample variance random variable $S^2 = (n-1)^{-1}\sum_i (X_i - \overline{X})^2$ given in Equation (6.4).

7.3.3.1 Where We Show $\overline{X}$ and S^2 Are Independent
Given that $\overline{X}$ is part of the definition of S^2, this result is a little surprising. As before, we may assume $\mu = 0$ w.l.o.g. We compute the covariance of $\overline{X}$ and $X_i - \overline{X}$, which is a typical term in S^2. Both have mean zero, so that the covariance is $E[\overline{X} \cdot (X_i - \overline{X})] = E[\overline{X}X_i - \overline{X}^2] = E[\overline{X}X_i] - E[\overline{X}^2]$. The term $E[\overline{X}^2] = \text{var}\overline{X} = \sigma^2/n$, since $\mu = 0$. Next,

$$\overline{X}X_i = \left(\frac{1}{n}X_i + \frac{1}{n}\sum_{j \neq i} X_j\right) \cdot X_i = \frac{1}{n}X_i^2 + \frac{1}{n}\sum_{j \neq i} X_j X_i.$$

Now $EX_i^2 = \text{var}(X_i) = \sigma^2$ since $\mu = 0$, and $E(X_j X_i) = 0$ by independence. Finally,

$$\text{cov}(\overline{X}, X_i - \overline{X}) = E[\overline{X}X_i] - \text{var}(\overline{X}) = \frac{1}{n}\sigma^2 + 0 - \frac{\sigma^2}{n} = 0.$$

This shows that those random variables are uncorrelated, not independent. However, since both $\overline{X}$ and $X_i - \overline{X}$ are linear combinations of a normal random sample, each is normal as well; see Equation (3.50). Hence, uncorrelated implies independence here. Since S^2 is made up of the n terms $X_i - \overline{X}$, each of which is independent of $\overline{X}$, then so must S^2 be independent of $\overline{X}$.

7.3.3.2 Where We Show That S^2 Scaled Is $\chi^2(n-1)$

To prove this result, we make use of Problem 4 and the following identity:

$$\sum (X_i - \overline{X})^2 = \sum [(X_i - \mu) - (\overline{X} - \mu)]^2$$
$$= \sum (X_i - \mu)^2 - 2(\overline{X} - \mu)\sum(X_i - \mu) + n(\overline{X} - \mu)^2$$
$$= \sum (X_i - \mu)^2 - n(\overline{X} - \mu)^2.$$

Rearranging and dividing through by σ^2, we have

$$\underbrace{\sum \frac{(X_i - \overline{X})^2}{\sigma^2}}_{\text{independent}} + \underbrace{\frac{(\overline{X} - \mu)^2}{\sigma^2/n}}_{\chi^2(1)} = \underbrace{\sum \left(\frac{X_i - \mu}{\sigma}\right)^2}_{\chi^2(n)}. \tag{7.12}$$

The first term is a scaled version of S^2, so it is independent of the second term that involves only $\overline{X}$. Thus, using the MGF for sums of independent χ^2 random variables, we see that their degrees of freedom add up; therefore, the first term in Equation (7.12) must be $\chi^2(n-1)$. Thus, we have shown

$$\boxed{\frac{n-1}{\sigma^2} S^2 \sim \chi^2(n-1).} \tag{7.13}$$

7.3.3.3 Where We Finally Derive the T PDF

It is easy to check that the statistic in Equation (7.11) is the same as

$$T_{n-1} = \frac{\overline{X} - \mu_0}{S/\sqrt{n}} = \frac{\frac{\overline{X} - \mu_0}{\sigma/\sqrt{n}}}{\sqrt{\frac{(n-1)S^2}{\sigma^2}/(n-1)}} = \frac{N(0,1)}{\sqrt{\frac{\chi^2(n-1)}{n-1}}}, \tag{7.14}$$

where the two random variables $N(0,1)$ and $\chi^2(n-1)$ are independent. Finding the PDF of T_{n-1} amounts to performing a bivariate change of variables and integrating out the second variable.

Introduce $Z \sim N(0,1)$ and $X \sim \chi^2(n-1)$, which are independent; hence,

$$f_{Z,X}(z,x) = \frac{1}{\sqrt{2\pi}}e^{-z^2/2} \cdot \frac{1}{\Gamma\left(\frac{n-1}{2}\right) 2^{\frac{n-1}{2}}} x^{\frac{n-1}{2}-1} e^{-x/2}, \tag{7.15}$$

using the definition of $\chi^2(p)$ in Appendix (B) with $p = n - 1$.

Define two new random variables S and T (where we use lower case for clarity):

$$
\left.\begin{array}{l}
s = s(z, x) = x \\[2mm]
t = t(z, x) = \dfrac{z}{\sqrt{x/(n-1)}}
\end{array}\right\} \quad 1\text{-}1 \quad
\left\{\begin{array}{l}
z = z(s, t) = t\sqrt{s/(n-1)}, \\[2mm]
x = x(s, t) = s,
\end{array}\right.
$$

which we may check are 1–1 over their domains.

Now the Jacobian of this transformation is given by

$$
J = \left\| \begin{array}{cc} \dfrac{\partial z(s, t)}{\partial s} & \dfrac{\partial z(s, t)}{\partial t} \\[3mm] \dfrac{\partial x(s, t)}{\partial s} & \dfrac{\partial x(x, t)}{\partial t} \end{array} \right\| = \left\| \begin{array}{cc} \dfrac{t}{2\sqrt{(n-1)s}} & \sqrt{\dfrac{s}{n-1}} \\[3mm] 1 & 0 \end{array} \right\| = -\sqrt{\dfrac{s}{n-1}}.
$$

Then using Equations (4.28) and (7.15),

$$
g_{S,T}(s, t) = \frac{1}{\sqrt{2\pi}} e^{-\frac{1}{2}\frac{t^2 s}{n-1}} \cdot \frac{1}{\Gamma\left(\frac{n-1}{2}\right) 2^{\frac{n-1}{2}}} s^{\frac{n-1}{2}-1} e^{-s/2} \cdot \left| -\sqrt{\frac{s}{n-1}} \right|.
$$

Using Mathematica to integrate out s to obtain the marginal of t, we have

$$
g_T(t) = \frac{\Gamma(n/2)}{\Gamma((n-1)/2) \sqrt{(n-1)\pi} \, (1 + t^2/(n-1))^{-n/2}} \sim T_{n-1},
$$

comparing $g_T(t)$ to the definition of T_p in Appendix B with $p = n - 1$.

Remarks: The T_p PDF is symmetric around 0. There are two special cases of note. When $p = 1$, the T_1 PDF is the Cauchy PDF; see Equation (4.26). And as $p \to \infty$, the PDF converges to the standard normal PDF.

7.3.4 The One-Sample T-test

We are now ready to find the critical region for the test given in Equation (7.10), where the variance σ^2 for both hypotheses is the same but unknown. Thus we need to find the maximum likelihood estimates of both (μ, σ^2) when μ is known (null hypothesis) or unknown (alternative hypothesis).

From Equation (7.6), the log-likelihood is

$$
\ell(\mu, \sigma^2 | \mathbf{x}) = -\frac{n}{2} \log 2\pi - \frac{n}{2} \log \sigma^2 - \frac{1}{2\sigma^2} \sum_{i=1}^{n} (x_i - \mu)^2. \tag{7.16}
$$

Under the alternative hypothesis, H_1, the MLEs are

$$
\hat{\mu}_1 = \bar{x} \quad \text{and} \quad \hat{\sigma}_1^2 = \frac{1}{n} \sum_{i=1}^{n} (x_i - \bar{x})^2,
$$

while $\mu = \mu_0$ is known under the null hypothesis, H_0. It is easy to see

$$
\hat{\mu}_0 = \mu_0 \quad \text{and} \quad \hat{\sigma}_0^2 = \frac{1}{n} \sum_{i=1}^{n} (x_i - \mu_0)^2.
$$

Denoting the constant $-\frac{n}{2}\log(2\pi)$ by c in Equation (7.16), we have

$$\ell_0 = c - \frac{n}{2}\log\left[\frac{1}{n}\sum(x_i - \mu_0)^2\right] - \frac{n}{2\sum(x_i - \mu_0)^2}\cdot\sum(x_i - \mu_0)^2$$

$$\ell_1 = c - \frac{n}{2}\log\left[\frac{1}{n}\sum(x_i - \bar{x})^2\right] - \frac{n}{2\sum(x_i - \bar{x})^2}\cdot\sum(x_i - \bar{x})^2.$$

The last terms in ℓ_0 and ℓ_1 are both $-n/2$; hence, the log-likelihood ratio is

$$\ell_0 - \ell_1 = 0 - \frac{n}{2}\log\left[\frac{\frac{1}{n}\sum(x_i - \mu_0)^2}{\frac{1}{n}\sum(x_i - \bar{x})^2}\right] - 0.$$

Letting $u = \sum(x_i - \mu_0)^2 / \sum(x_i - \bar{x})^2$, the critical region is determined by

$$\ell_0 - \ell_1 = -\frac{n}{2}\cdot\log(u) < k \quad\Longleftrightarrow\quad u > k'.$$

We can re-write the numerator of u as

$$\sum(x_i - \mu_0)^2 = \sum[(x_i - \bar{x}) - (\mu_0 - \bar{x})]^2 \quad\text{(add and subtract } \bar{x}\text{)}$$

$$= \sum[(x_i - \bar{x})^2 - 2(x_i - \bar{x})(\mu_0 - \bar{x}) + (\mu_0 - \bar{x})^2]$$

$$= \sum(x_i - \bar{x})^2 - 0 + n(\bar{x} - \mu_0)^2,$$

since $\sum(x_i - \bar{x}) = 0$. Noting $T_{n-1}^2 = (\bar{x} - \mu_0)^2/(s^2/n)$ from Equation (7.11),

$$T_{n-1}^2 = \frac{(\bar{x} - \mu_0)^2\, n}{\frac{1}{n-1}\sum(x_i - \bar{x})^2} = \frac{n(n-1)\,(\bar{x} - \mu_0)^2}{\sum(x_i - \bar{x})^2},$$

we have

$$u = \frac{\sum(x_i - \mu_0)^2}{\sum(x_i - \bar{x})^2} = \frac{\sum(x_i - \bar{x})^2 + n(\bar{x} - \mu_0)^2}{\sum(x_i - \bar{x})^2}$$

$$= 1 + \frac{n(\bar{x} - \mu_0)^2}{\sum(x_i - \bar{x})^2} = 1 + (n-1)^{-1}\cdot T_{n-1}^2 > k'$$

$$\boxed{\Longleftrightarrow \qquad T_{n-1} \notin (-a, a), \quad \text{we reject } H_0,} \tag{7.17}$$

for $a = \sqrt{(k' - 1)\cdot(n - 1)}$. Equation (7.17) gives an equal-tail-area critical region, which is appropriate since the PDF of T_{n-1} is symmetric around 0.

How different are the widths of the "acceptance regions" $(-a, a)$ for the Z-test and the T-test? The Table below gives the values of a for a 95% confidence level test for several values of n. In the past, statisticians reverted to Z-tests when the sample size exceeded 50, but with complete T_{n-1} tables available in **R**, there is no need. It should be clear why this result began the field of **small-sample statistics**.

7.3.5 Example

Colton (1974) measured the heart rates of ten male medical students: (58, 59, 62, 62, 65, 72, 72, 77, 77, 83), for which $\bar{x} = 68.7$ and $s^2 = 75.122$. He wished to compare it to a published standard of 72 beats per minute.

$$H_0\colon \mu = 72 \quad\text{versus}\quad H_1\colon \mu \neq 72.$$

Table 7.2 Upper critical values for the two-sided Z-test versus T-test.

n	$df = n-1$	$z_{0.975}$	$t_{0.975,n-1}$	% wider
2	1	1.96	12.706	548.3%
3	2	1.96	4.303	119.5%
4	3	1.96	3.182	62.4%
5	4	1.96	2.776	41.7%
7	6	1.96	2.447	24.8%
10	9	1.96	2.262	15.4%
15	14	1.96	2.145	9.4%
20	19	1.96	2.093	6.8%
30	29	1.96	2.045	4.4%
50	49	1.96	2.010	2.5%
100	99	1.96	1.984	1.2%
200	199	1.96	1.972	0.6%
500	499	1.96	1.965	0.2%

Computing the test statistic in Equation (7.11)

$$T_{10-1} = \frac{68.7 - 72}{\sqrt{75.122/10}} = -1.204.$$

From Table 7.2, we see the test statistic is in the 95% interval $(-2.262, 2.262)$; hence, we do not reject the null hypothesis.

7.3.6 Other T-tests

Fisher was supportive of Gosset's efforts to publish his paper, as Fisher had discovered other instances where the T_p PDF was useful.

7.3.6.1 Paired T-test

When evaluating the efficacy of a new procedure or treatment, one must usually allocate individuals randomly to the control group as well as the treatment group. The ethics of such a design are always in mind, as the patients who randomly receive the placebo or the old treatment do not always agree with the design, and may choose not to participate.

In some situations, it is possible to give both treatments to each individual, again, in random order. This so-called **self-control design** has many advantages: primarily, fewer patients need to be recruited to achieve a good decision, and recruiting may also be easier.

The data in r.v. form are (X_i, Y_i), where the standard/placebo treatment outcome is recorded as X_i. The alternate/new treatment outcome is recorded as Y_i. The analysis involves simply computing the difference $D_i = Y_i - X_i$ and testing the hypotheses

$$H_0: \mu_D = 0 \quad \text{versus} \quad H_1: \mu_D \neq 0, \quad \text{assuming } D \sim N(0, \sigma^2), \tag{7.18}$$

where σ^2 is unknown. Notice that we do not perform a one-sided alternative, as the new treatment might prove inferior. The analysis is precisely that in Section 7.3.4 with data $\{d_1, d_2, \ldots, d_n\}$ and with $\mu_0 = 0$.

We should mention that occasionally pairs of individuals are "matched" on a list of other co-variates. Then each pair is randomly assigned a treatment, and the difference recorded. This procedure can represent an improvement compared to the true two-sample procedure discussed in the next section, but it is highly dependent upon the success of matching. Matching is seldom close to ideal.

7.3.6.2 Two-Sample *T*-test

Many experiments do not realistically permit both treatments to be given to each individual. For example, imagine running an experiment that compares knee replacement surgery verses an injection in the knee!

Patients are allocated randomly to the two treatment groups, where $n_X + n_Y = n$. The allocation need not be 50–50. The data are recorded in r.v. form as $X_1, \ldots, X_{n_X}$ and $Y_1, \ldots, Y_{n_Y}$. The hypotheses we wish to test are

$$H_0: \mu_X = \mu_Y \quad \text{versus} \quad H_1: \mu_X \neq \mu_Y, \tag{7.19}$$

assuming $X_i \sim N(\mu_X, \sigma^2)$ and $Y_i \sim N(\mu_Y, \sigma^2)$ under the alternative hypothesis, that is, with unknown but common variance. Under the null hypothesis, the two means are assumed to be equal.

Since all the data are independent, we have that

$$\overline{Y} - \overline{X} \sim N\left(\mu_Y - \mu_X, \frac{\sigma^2}{n_X} + \frac{\sigma^2}{n_Y}\right);$$

hence, the generalization of the test statistic in Equation (7.14) is

$$T_{n-2} = \frac{(\overline{Y} - \overline{X}) - (\mu_Y - \mu_X)}{\sqrt{S_P^2 \left(\frac{1}{n_X} + \frac{1}{n_Y}\right)}} \tag{7.20}$$

where S_P^2 is the so-called **pooled variance** given by

$$S_P^2 = \frac{\sum (X_i - \overline{X})^2 + \sum (Y_i - \overline{Y})^2}{n_X + n_Y - 2}. \tag{7.21}$$

We note that this pooled estimator is unbiased for σ^2. Each sum scaled by σ^2 is $\chi^2(n_X - 1)$ and $\chi^2(n_Y - 1)$, so the pooled sum is $\chi^2(n_X + n_Y - 2)$ or $\chi^2(n - 2)$ because of independence.

7.3.6.3 Example Two-Sample *T*-test: Lord Rayleigh's Data

Review the exploratory analysis of Lord Rayleigh's experiment in Section 1.1.2. Of the 24 measurements he took over the 19 months he collected data, 14 are listed as having an origin that is "air" while 10 are listed as "chemical" in origin. We wish to test the null hypothesis that the means are the same. This is an example of Equation (7.19).

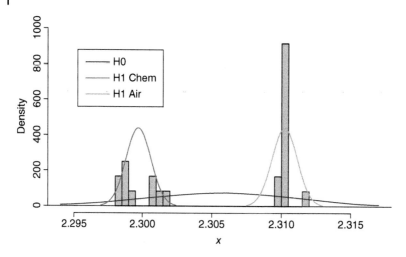

Figure 7.7 Fits to Lord Rayleigh's data under the null and alternative hypotheses. A Nobel prize was awarded for understanding this diagram.

Under the null hypothesis, we treat the data as one normal sample of size $n = 24$, for which $\bar{x} = 2.3058$ and $s = 0.00537$; see Figure 7.7. The pooled estimate of common variance is 8.26331×10^{-7} from Equation (7.21), while $\bar{x} = 2.2997$ and $\bar{y} = 2.3102$. Thus the test statistic is

$$t_{24-2} = \frac{(2.3102 - 2.2997) - 0}{\sqrt{8.26331 \times 10^{-7}\left(\frac{1}{14} + \frac{1}{10}\right)}} = 27.96,$$

which far exceeds the 5% or 1% test thresholds of $t_{.975,22} = 2.074$ or $t_{.995,22} = 2.819$. Thus we reject the null hypothesis that the origin of the nitrogen sample does not matter – the air samples are heavier. Examining the bimodal feature of the chemical samples in Figure 7.7 suggests that perhaps there is something further to be understood about those samples. The other variables recorded were the method of extraction and the purifying agent. Finally, compare the visual impact of Figure 7.7 to Figure 1.3(Left), where the bins were carefully selected to make the histogram appear unimodal.

7.4 Reporting Results: *p*-values and Power

Looking back to the beginning of this chapter and the JAMA instructions about reporting results, we see mention of type I error and *p*-values. Picking a level α test determines the boundary of the critical region. Given the experimental data lands in the critical region, we can report rejecting or failing to reject the null hypothesis, respectively, at the pre-determined level.

The *p*-value complements the type I error by giving the minimum value of α where the null hypothesis **would not be rejected**. For this choice of α, the data would be located on the boundary of the critical region; see below.

7.4.1 Example When the Null Hypothesis Is Rejected

Consider the alternative $\mu \neq 10$ for the shifted normal model described in Section 7.2.2 with $n = 16$ and $\sigma = 5$ at the $\alpha = 5\%$ level. The two-sided critical region is

$$\left| \frac{\overline{X} - \mu_0}{\sigma/\sqrt{n}} \right| > z_{(1-\alpha/2)} \quad \text{or} \quad \overline{X} \notin \left(10 \mp 1.96 \cdot \frac{5}{\sqrt{16}} \right) = (7.55, 12.45).$$

Suppose the experimental outcome is $\overline{x} = 5.0$, and we report rejecting the null hypothesis. If the critical region had been $C = (5.0, 15.0)^c$, then the test statistic would have been on the boundary. Working backwards, we see

$$\frac{\overline{x} - \mu_0}{\sigma/\sqrt{n}} = \frac{5 - 10}{5/\sqrt{16}} = -4.0,$$

which is the 3.167×10^{-5} percentile. Choosing α to be twice this would put $\overline{x} = 5$ on the boundary of C. Thus we would report p-value as $p = 6.334 \times 10^{-5}$.

To restate this, choosing $\alpha = 6.334 \times 10^{-5}$, gives $z_{(1-\alpha/2)} = 4.0$ and the critical region $C = (5.0, 15.0)^c$. This experiment would reject the null hypothesis at the 1% level as well. Sometimes the phrase **highly significant** is used in this situation. The p-value gives the probability we would have seen **this result or one more extreme, assuming the null hypothesis is true**.

Lord Rayleigh example (continued): The T-statistic from the experiment in Section 7.3.6.3 was 27.956 on 22 degrees of freedom. Thus the p-value (two-sided) is found using the **R** function pt as

$$p\text{-value} = 2 * \mathtt{pt}(-27.956, 22) = 1.10 \times 10^{-18},$$

which is highly significant. The origin of the nitrogen sample matters.

7.4.2 When the Null Hypothesis is Not Rejected

If the first result in the previous section had been $\overline{x} = 12.0$ instead of 5.0, then we would report *failing to reject* the null hypothesis. The p-value is computed the same way as before:

$$\frac{\overline{x} - \mu_0}{\sigma/\sqrt{n}} = \frac{12.0 - 10}{5/\sqrt{16}} = 1.6,$$

which is the 94.52 percentile. Thus choosing $\alpha = 2 \cdot (1 - .9452) = 10.96\%$ would give the interval $(10 \pm z_{(1-.0548)} \cdot 1.25)$ or $(8.0, 12.0)$ and $\overline{x}$ on the boundary. While the odds of seeing an experimental result this extreme is only about 1 in 9 or 10, we would still fail to reject at the 5% level.

Running an expensive experiment that concluded with a p-value of 5.02% would be very stressful. Should we continue to collect data in hopes that the p-value falls below 5%? The JAMA instructions would frown upon such a decision.

7.4.3 The Power Function

The power function extends calculations such as the p-value to the entire range of possible values of a parameter θ. Specifically, given a critical region C at a specified α level,

$$\beta(\theta) = P(\text{reject } H_0 \mid \theta \text{ is true parameter}). \tag{7.22}$$

Notice the type I error is $\beta(\theta_0)$. For $\theta \neq \theta_0$, $\beta(\theta)$ gives the probability of rejecting H_0 correctly. A type II error in this case would be $1 - \beta(\theta)$; see Equation (7.8).

Continuing with our example in Section 7.4.1 testing $\mu \neq \mu_0$ with $\mu_0 = 10$ and $\sigma = 5$, and working with the complement of C:

$$\beta(\mu) = 1 - Pr\left(\mu_0 - \frac{1.96\sigma}{\sqrt{n}} < \overline{X} < \mu_0 + \frac{1.96\sigma}{\sqrt{n}} \;\middle|\; \mu_X = \mu \right)$$

$$= 1 - Pr\left(\mu_0 - \mu - \frac{1.96\sigma}{\sqrt{n}} < \overline{X} - \mu < \mu_0 - \mu + \frac{1.96\sigma}{\sqrt{n}} \right)$$

$$= 1 - Pr\left(\frac{\mu_0 - \mu}{\sigma/\sqrt{n}} - 1.96 < \frac{\overline{X} - \mu}{\sigma/\sqrt{n}} < \frac{\mu_0 - \mu}{\sigma/\sqrt{n}} + 1.96 \right),$$

where we have manipulated the inequalities so that $\overline{X} \sim N(\mu, \sigma^2/n)$ is standardized. The power function is displayed in the left frame of Figure 7.8. The power at $\mu = 7.25$ is 59.5%, and 89.3% at $\mu = 14.0$. If we were interested in detecting a shift in the mean from 10 to 7.25, the sample size $n = 16$ would be insufficient to reliably reject the null hypothesis. A shift to 14, however, would be detected with high probability.

The right frame of Figure 7.8 shows the influence of increasing the sample size on the power function. In many situations, specifying both the type I error (e.g., $\alpha = 1\%$) and the power function at a value of μ where detection is desired will give two equations that determine not only the critical region but also the sample size.

For example, if we choose $\alpha = 5\%$ and $\beta(12) = 0.90\%$ then the acceptance region is $C^c = (8.79, 11.21)$ and $n = 65.67$. If we have the resources to choose $\alpha = 1\%$ and

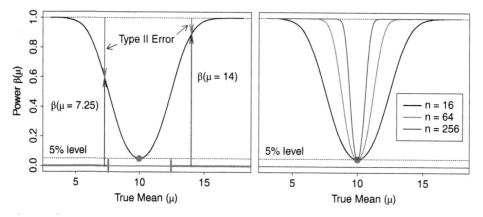

Figure 7.8 (Left) Power function with $n = 16$. The critical region is shown in gray, along with two values of the power function at $\mu = 7.25$ and 14.0. Their complements are examples of type II errors. (Right) Effect of sample size on the power function.

$\beta(12) = 0.98\%$, then the critical region is $C^c = (8.89, 11.11)$ and $n = 133.96$; see Problem 6. In practice, we would round upwards and choose $n = 66$ and $n = 134$ and have slightly more power.

7.5 Multiple Testing and the Bonferroni Correction

A research paper typically reports on more than one hypothesis test result. If each test is conducted at a significance level α (probability of a type I error), what can we say about the overall error probability?

Suppose we conducted two independent experiments, each at the 5% level. Suppose further that the null hypotheses were in fact true for both. Then the probability of erroneously rejecting the null hypothesis is 5% for each, but

$$1 - (1 - .05)(1 - .05) = 1 - 0.95^2 = 0.0975 \tag{7.23}$$

for both. This follows a binomial experiment, $X \sim \text{Binom}(2, p)$, where the probability $p = 0.95$ of correctly *not rejecting the null hypothesis* in two independent trials. $X = 2$ is the event that both hypotheses are not rejected; hence, the probability in Equation (7.23) is $p_X(0) + p_X(1)$, where either one or both hypotheses are rejected. Note $p_X(0) = 0.05^2 = 0.0025$, $p_X(1) = 2 \cdot 0.95 \cdot 0.05 = 0.0950$, and $p_X(2) = 0.95^2 = 0.9025$.

Bonferroni simply observed that if we wish to guarantee an *overall type I error rate* at the α-level, then run each of k independent hypothesis tests at the $\alpha' = \alpha/k$ level. If so, and all the null hypotheses are true, the probability of correctly failing to reject all k is given by $(1 - \alpha')^k$; hence, the probability of erroneously rejecting one or more is given by

$$\boxed{P(\text{overall type I error}) = 1 - \left(1 - \frac{\alpha}{k}\right)^k = \alpha - \frac{k-1}{2k}\alpha^2 + \cdots \leq \alpha.} \tag{7.24}$$

For large k, this can be quite conservative, i.e., the true type I error rate is likely to be much less than 5% and the corresponding powers reduced as a consequence. Note the similarity to the parallel reliability model in Section 2.6.2. The Bonferroni method satisfies the JAMA requirements for handling multiple comparisons given just before Section 7.1.

Problems

7.1 Show that the sum of n i.i.d. negative exponential r.v.s with parameter β has a *Gamma(n, β)* exactly. Hint: use the MGF technique of Section 5.4.1.

7.2 Show that among the contiguous intervals (a, b) containing 95% of the probability for our example in Section 7.3.1.1, a necessary condition to minimize the width of the interval, $b - a$, is that the sampling density must be equal at a and b. This result holds generally if the sampling density is unimodal and monotone on either side of the mode.

7.3 Suppose the distribution of the r.v. used to find the various intervals (a, b) in Section 7.3.1.2 was not only unimodal and monotone on either side of the mode, but was also symmetric. Show that the equal-tail-area and narrowest-width intervals are identical.

7.4 Using the MGF technique, show that

$$Y = \sum_{i=1}^{n} Z_i^2 \sim \chi^2(n),$$

where Z_i is a random sample from the $N(0, 1)$ PDF.

7.5 Using the MGF technique, show that if $U = S + T$, where (a) $U \sim \chi^2(n)$; (b) $T \sim \chi^2(1)$; and (c) S and T are independent, then $S \sim \chi^2(n-1)$.

7.6 Verify the critical regions and sample sizes for the example at the end of Section 7.4.3.

7.7 The 1969 military draft lottery numbers are shown in Figure 7.9. Individuals with numbers over 195 were not drafted. There appears to be a non-random downward trend. For example, only five birthdays in December had a lottery number above average. Run a two-sample T-test in **R** with the command `t.test(x[1:183], x[184:366])` and show the p-value is 5×10^{-5}. What do we conclude? Perform a simulation using data from `x = sample(366)` to verify the T-distribution is the appropriate test statistic.

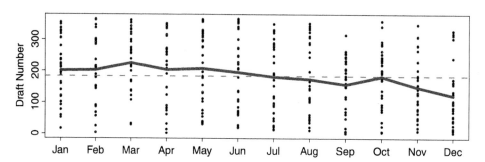

Figure 7.9 Draft lottery numbers 1–366 by month. The monthly average is the blue line. The overall average of 183.5 is the red dotted line.

8

Confidence Intervals and Other Hypothesis Tests

In this chapter we introduce the confidence interval (CI), which provides an additional way of reporting the result of a hypothesis test about a parameter. Then we apply these ideas to tests and models for the family of Pearson's goodness-of-fit χ^2-tests, the correlation coefficient, linear regression curve fitting, and, finally, for the analysis of variance (ANOVA), which tests the simultaneous equality of more than two normal means (thus extending the two-sample T-test). There is a strong and growing sentiment that experimental results should always include confidence intervals, rather than p-values by themselves. At times we take a more informal approach to our development than the formal likelihood ratio tests.

8.1 Confidence Intervals

The hypothesis test reports the decision either to **reject the null hypothesis** or to **fail to reject the null hypothesis**. The latter use of the double negative is intentional. It is quite difficult to **prove** the null hypothesis. Almost any alternative will have a mean that isn't exactly the same as μ_0, and with sufficient sample size, it can be "detected." Whether the difference has any practical importance is determined by the experts requesting the experiment.

The same test statistics can be used to construct and report a more informative summary, called the **confidence interval (CI)**. The confidence interval gives a range of **the most likely values of the parameter**. If the level of the test is 5%, the corresponding quantity is the 95% confidence interval. We show how to construct several CIs in the following sections.

8.1.1 Confidence Interval for μ: Normal Data, σ^2 Known

The acceptance region (C^c) for testing $H_0 : \mu = \mu_0$ versus $H_1 : \mu \neq \mu_0$ is

$$z_{0.025} \leq \frac{\overline{X} - \mu_0}{\sigma/\sqrt{n}} \leq z_{0.975}, \tag{8.1}$$

Statistics: A Concise Mathematical Introduction for Students, Scientists, and Engineers, First Edition. David W. Scott.
© 2020 John Wiley & Sons Ltd. Published 2020 by John Wiley & Sons Ltd.

a level 5% test. These two inequalities can be turned inside out to yield

$$\overline{X} + z_{0.025}\frac{\sigma}{\sqrt{n}} \leq \mu_0 \leq \overline{X} + z_{0.975}\frac{\sigma}{\sqrt{n}} \tag{8.2}$$

or, equivalently, for a general level $1 - \alpha$ confidence interval,

$$\mu_0 \in \overline{X} \mp z_{\left(1-\frac{\alpha}{2}\right)}\frac{\sigma}{\sqrt{n}}, \quad \textbf{with confidence } 100(1 - \alpha)\%. \tag{8.3}$$

The interpretation of a confidence interval requires some discussion. There is no doubt in the probability statement in Equation (8.1). There is also no doubt that the re-expressed version in Equation (8.2) is algebraically equivalent. However, as a probability statement, the fixed (non-random) parameter μ_0 is either in the **random interval** or not. Therefore, the probability is either 0% or 100%, not 95% for a particular experiment.

However, if 100 experimenters use this confidence interval and H_0 is true, we expect 95 of their reported CIs to include μ_0, while 5 will not. Suppose the reported CIs while testing $H_0 \colon \mu_0 = 0$ were

$$\mu_0 \in (2.5, 4.7) \qquad \text{versus} \qquad \mu_0 \in (-0.6, 1.6).$$

We know that $\bar{x} = 3.6$ and 0.5 (the midpoints), respectively, and that the decisions were to reject and accept, respectively. However, we also gain more quantitative information about the probable values of μ_0 in addition to the decision. A small simulation study illustrates these ideas in the next section.

8.1.2   Confidence Interval for μ: σ^2 Unknown

The acceptance region (C^c) for testing $H_0 \colon \mu = \mu_0$ versus $H_1 \colon \mu \neq \mu_0$ at the $1 - \alpha$ confidence level is

$$t_{\alpha/2,n-1} \leq \frac{\overline{X} - \mu_0}{S/\sqrt{n}} \leq t_{(1-\alpha/2),n-1}. \tag{8.4}$$

These two inequalities can be turned inside to yield

$$\mu_0 \in \overline{X} \mp t_{\left(1-\frac{\alpha}{2},n-1\right)}\frac{S}{\sqrt{n}}, \quad \textbf{with confidence } 100(1 - \alpha)\%. \tag{8.5}$$

It is straightforward to construct confidence intervals for paired and two-sample T-tests. The details are left to the reader.

Example: Suppose we collect a small sample of size $n = 16$ from a standard normal PDF, and that the null hypothesis $\mu = 0$ is true. Following Equation (8.5), one hundred 95% confidence intervals were generated and are plotted in the upper left frame of Figure 8.1. Six of the 100 confidence intervals, shown in red, did not include the true value of μ. The bottom left frame shows these same confidence intervals sorted for clarity. Notice three of the confidence intervals were below zero and three were above zero. In the right frames, 100 samples of size $n = 16$ were generated from a $N(0.7495, 1)$ PDF, which was chosen so that

$$\text{Prob}\left(\overline{X} - t_{0.975,15}\frac{S}{\sqrt{n}} < 0\right) = 20\%. \tag{8.6}$$

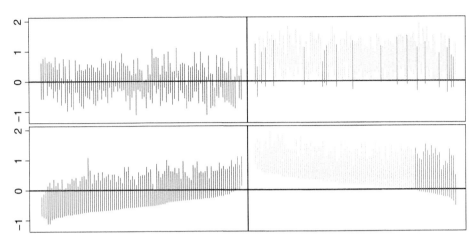

Figure 8.1 (Left frames) A simulation study showing 100 95% confidence intervals for samples of size $n = 16$ from the $N(0, 1)$ PDF, where the intervals that fail to include the true value of $\mu_0 = 0$ are shown in red. The bottom left frame shows the same 100 confidence intervals sorted for clarity. (Right frames) A simulation study from the alternative hypothesis PDF $N(0.7495, 1)$, chosen so that the power is 80%. In fact, 21 of the 100 CIs incorrectly include the null hypothesis mean 0.

Thus there is an 80% chance that the confidence interval would confirm that the alternative hypothesis is true. In this simulation, 21 of the 100 confidence intervals (again in red) incorrectly include $\mu = 0$. The value $\mu_1 = 0.7495$ was determined using the fact that the T statistic when the numerator is $N(\mu, 1)$ rather than $N(0, 1)$ follows a *non-central T-distribution*; see Problem 2.

8.1.3 Confidence Intervals and p-values

In hypothesis testing, the p-value gives the test level where the observed test statistic is on the boundary of the critical region, C. If we choose α to be the same as the p-value, then the confidence interval for testing $\mu_0 = 0$ will have the form

$$\mu_0 \in (0, 2.2),$$

or, alternatively, $\mu_0 \in (-2.2, 0)$, depending upon the sign of $\bar{x}$.

If the p-value had been 28.4%, then these would have been 71.6% confidence intervals. The decision would have been *fail to reject*.

If the p-value had been 0.028%, then these would have been 99.972% confidence intervals. The decision would have been *to reject* H_0. However, a p-value gives us only one piece of information, while a CI gives us two, since the CI also conveys information about the power of the test. Again, many statisticians believe p-values should not be reported alone.

8.2 Hypotheses About the Variance and the *F*-Distribution

In this section, we assume we have a random sample of normal data from one or two populations. Instead of hypotheses about the mean(s), we consider hypotheses about the variance(s). We begin by introducing the F-distribution.

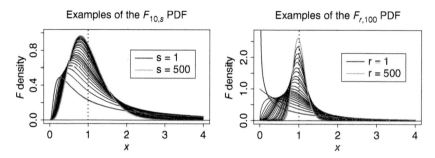

Figure 8.2 (Left) Examples of the $F_{10,s}$ PDF for $1 \leq s \leq 500$. (Right) Examples of the $F_{r,100}$ PDF for $1 \leq r \leq 500$.

8.2.1 The F-Distribution

Snedecor's F-distribution is the key PDF when comparing two independent sample variances. Rather than looking at the differences of two χ^2 random variables, the F-distribution is defined by their ratio

$$
\boxed{\text{Definition of the } F \text{ random variable:} \quad F_{r,s} = \frac{\chi_r^2/r}{\chi_s^2/s}.}
\tag{8.7}
$$

In Figure 8.2, we show some examples of the $F_{r,s}$ PDF.

8.2.2 Hypotheses About the Value of the Variance

We begin by considering two simple hypotheses for our $N(\mu, \sigma^2)$ sample:

$$
H_0 : \sigma^2 = \sigma_0^2 \quad \text{versus} \quad H_1 : \sigma^2 = \sigma_1^2 \neq \sigma_0^2.
$$

Thus the log-likelihoods under H_0 and H_1 are

$$
\ell(\sigma_0^2 | \mathbf{x}, \mu) = -\frac{n}{2} \log 2\pi - \frac{n}{2} \log \sigma_0^2 - \frac{1}{2\sigma_0^2} \sum_{i=1}^{n} (x_i - \mu)^2 \quad \text{and}
$$

$$
\ell(\sigma_1^2 | \mathbf{x}, \mu) = -\frac{n}{2} \log 2\pi - \frac{n}{2} \log \sigma_1^2 - \frac{1}{2\sigma_1^2} \sum_{i=1}^{n} (x_i - \mu)^2.
$$

Neither hypothesis specifies the value of the mean, μ, so we replace it with the maximum likelihood estimate, which is $\bar{x}$. Hence, the form of the critical region is given by the difference $\ell(\sigma_0^2 | \mathbf{x}, \bar{x}) - \ell(\sigma_1^2 | \mathbf{x}, \bar{x})$, which equals

$$
\frac{n}{2} \log \frac{\sigma_1^2}{\sigma_0^2} + \frac{\sum (x_i - \bar{x})^2}{2} \left(\frac{1}{\sigma_1^2} - \frac{1}{\sigma_0^2} \right) \leq \log k \quad \text{or}
$$

$$
\sum (x_i - \bar{x})^2 \left(\frac{\sigma_0^2 - \sigma_1^2}{\sigma_0^2 \sigma_1^2} \right) \leq k' - n \log \frac{\sigma_1^2}{\sigma_0^2}.
$$

Thus the test involves the sample variance, S^2, with the inequality depending upon the sign of $\sigma_0^2 - \sigma_1^2$:

$$\text{reject } H_0 \text{ when } \begin{cases} S^2 \geq k'' & \text{if } \sigma_1^2 > \sigma_0^2 \\ S^2 \leq k'' & \text{if } \sigma_1^2 < \sigma_0^2. \end{cases}$$

The critical region is determined by recalling that under the null hypothesis,

$$\frac{(n-1)S^2}{\sigma_0^2} \sim \chi_{n-1}^2 ; \qquad \text{hence, an } \alpha \text{ -level test is}$$

$$\boxed{\text{reject } H_0 \text{ when } \quad S^2 \gtrless \chi_{n-1}^2(1-\alpha)\,\frac{\sigma_0^2}{n-1} \qquad \text{if } \sigma_1^2 \gtrless \sigma_0^2,} \qquad (8.8)$$

where $\chi_{n-1}^2(1-\alpha)$ denotes the $100(1-\alpha)$ percentile.

Example 8.1   If $\sigma_0 = 1$ and $\sigma_1 > 1$, then we reject H_0 if $S^2 > 1.517$ or $S^2 > 1.245$ for $n = 25$ and 100, respectively, at the 5% level. This is equivalent to the inequalities $S > 1.232$ and $S > 1.116$ involving the sample standard deviation.

8.2.3   Confidence Interval for the Variance

Using the equal-tail area option, let

$$a = \chi_{n-1}^2(\alpha/2) \qquad \text{and} \qquad b = \chi_{n-1}^2(1-\alpha/2). \qquad (8.9)$$

Then the $100(1-\alpha)\%$ confidence intervals for σ^2 and σ are given by

$$a \leq \frac{(n-1)S^2}{\sigma^2} \leq b \qquad \Longleftrightarrow \qquad \frac{1}{b} \leq \frac{\sigma^2}{(n-1)S^2} \leq \frac{1}{a} \qquad \text{or}$$

$$\boxed{\frac{(n-1)S^2}{b} \leq \sigma^2 \leq \frac{(n-1)S^2}{a} \qquad \text{or} \qquad \frac{S\sqrt{n-1}}{\sqrt{b}} \leq \sigma \leq \frac{S\sqrt{n-1}}{\sqrt{a}}.} \qquad (8.10)$$

Example 8.2   If $S^2 = 1.4$ ($S = 1.183$), then 95% confidence intervals for σ when $n = 25$ and 100 are $(0.909, 1.600)$ and $(1.034, 1.367)$, respectively. Compare to Example 8.1 above in Section 8.2.2.

8.2.4   Two-Sided Alternative for Testing $\sigma^2 = \sigma_0^2$

For testing the hypotheses

$$H_0 : \sigma^2 = \sigma_0^2 \quad \text{versus} \quad H_1 : \sigma^2 \neq \sigma_0^2,$$

we compute the statistic

$$\boxed{\frac{(n-1)S^2}{\sigma_0^2} \sim \chi_{n-1}^2 \quad \text{under } H_0 ;} \qquad (8.11)$$

if the statistic falls in the interval (a, b), where a and b were defined in Equation (8.9), we fail to reject H_0. Otherwise, we reject H_0.

Example 8.3 Following Example 8.2 in the previous section, if $S^2 = 1.4$, then the test statistics for $n = 25$ and $n = 100$ are 33.60 and 138.60, respectively. The intervals (a, b) are $(12.40, 39.36)$ and $(73.36, 128.42)$. Thus we fail to reject $H_0 : \sigma_0^2 = 1$ when $n = 25$, but reject H_0 when $n = 100$.

8.3 Pearson's Chi-Squared Tests

Some of the most influential and lasting of Karl Pearson's works are several tests involving the multinomial PDF, which is introduced in the next section. These can be used to evaluate the adequacy of a parametric model. Other versions test the independence or dependence of two categorical variables arranged in a table. These all turn out to follow a χ^2 test statistic. In a rare mistake, Pearson did not choose the correct degrees of freedom for the χ^2 r.v. Fisher wrote a paper correcting the error, which Pearson declined to publish in Biometrika. This, and perhaps other events, led to a lifetime of animosity between the two giants of statistics, an unfortunate historical fact.

8.3.1 The Multinomial PMF

The multinomial PMF is a generalization of the binomial PMF, but with K categories instead of only two. The vector $\mathbf{p} = (p_1, p_2, \ldots, p_K)$ assigns probabilities to the K groups; of course, $\sum p_k = 1$. Each trial results in membership to one (and only one) of the K groups. After n independent trials, the number of observations in the kth group is denoted by x_k, so that $\sum x_k = n$. Any single sequence of n trials with counts $\mathbf{x} = (x_1, x_2, \ldots, x_K)$ has probability $p_1^{x_1} p_2^{x_2} \cdots p_K^{x_K}$. With two groups, the number of orderings of the n sequences is given by the binomial coefficient. With K groups, the number of orderings is given by the multinomial coefficient, which was introduced in Section 2.3.5. Thus, the multinomial PMF is given by

$$P_X(n, \mathbf{p}) = P(\mathbf{X} = \mathbf{x}) = \binom{n}{x_1, x_2, \ldots, x_K} p_1^{x_1} p_2^{x_2} \cdots p_K^{x_K}. \tag{8.12}$$

We denote this PMF by $\mathbf{X} \sim \text{MultiNom}(n, \mathbf{p})$.

It is easy to see that $EX_k = np_k$ and $\text{var}X_k = np_k(1 - p_k)$ since a measurement is either in the kth group or not (hence, binomial-like). It may also be shown that $\text{cov}(X_k, X_\ell) = -np_k p_\ell$.

The multinomial PMF may be used to model the counts on the roll of a die, or the bin counts in a histogram, for example. Pearson's several χ^2-tests correspond to various assumptions about the probability vector $\mathbf{p}$.

8.3.2 Goodness-of-Fit (GoF) Tests

Are our data normal? Or perhaps some other PDF? When modeling data with a parametric density, $f(x|\theta)$, this question should be step one. A histogram (in probability form) can provide not only graphical evidence, but also the bin counts to use in Pearson's goodness-of-fit (GoF) test, which we now describe.

We begin by constructing a histogram of the random sample choosing a finite number, m, of equally spaced bins $B_k = (kh, (k+1)h)$ that contain all the data. The associated bin counts are $\{v_k\}$, where v_k is the number of the n random samples contained in the interval B_k. In Pearson's jargon, these give rise to the so-called **observed counts** and **expected counts**

$$o_i = v_i$$
$$e_i = n \times \int_{B_i} f(x|\theta) \, dx.$$

The m bins span an interval (a, b) and the m bin counts $\{v_k\}$ follow a multinomial PDF. A hypothesis test for a multinomial PDF with m categories might be that for a given vector of probabilities $\mathbf{p}_0 = (p_1, p_2, \ldots, p_m)$:

$$H_0 : \mathbf{p} = \mathbf{p}_0 \quad \text{versus} \quad H_1 : \mathbf{p} \neq \mathbf{p}_0.$$

Rather than using the LR approach, we begin with a binomial test ($m = 2$) and then generalize to the full m-category multinomial case.

8.3.3 Two-Category Binomial Case

When there are only $m = 2$ categories, we may model the two bin counts as X and $n - X$, where $X \sim \text{Binom}(n, p)$. Here, p and $1 - p$ are the two bin probabilities.

If both n and np are sufficiently large, then the central limit theorem holds, and we can construct an approximate Z-test for the hypotheses

$$H_0 : p = p_0 \quad \text{versus} \quad H_1 : p \neq p_0.$$

At the α-level, the test is constructed by using the statistic

$$Z = \frac{X - np_0}{\sqrt{np_0(1 - p_0)}} \quad \text{and rejecting } H_0 \text{ when} \quad |Z| > z_{1-\alpha/2} \; ; \qquad (8.13)$$

here, Z simply measures the distance in standard units that the data count is from the null hypothesis. Equivalently, we may reject H_0 when

$$Z^2 > z_{1-\alpha/2}^2 \quad \text{or} \quad Z^2 > \chi_1^2(1 - \alpha), \qquad (8.14)$$

since $Z^2 \sim \chi_1^2$. The two-tailed Z-test becomes a one-tailed Z^2 or χ_1^2 test. Thus we see we now have a χ^2 statistic.

Algebraically, it is easy to check that we may rewrite Z^2 as

$$Z^2 = \frac{(X - np_0)^2}{np_0(1 - p_0)} = \frac{(X - np_0)^2}{np_0} + \frac{(X - np_0)^2}{n(1 - p_0)}. \qquad (8.15)$$

Now the bin count in the second bin is $n - X$, which has expectation $n - np_0 = n(1 - p_0)$ under H_0; hence, the difference between the bin count and expectation for the second bin equals

$$(n - X) - (n(1 - p_0)) = -X + np_0 = -(X - np_0). \qquad (8.16)$$

If we adopt the multinomial PDF $M(n; p_1, p_2)$ with $m = 2$ to represent the Binom(n, p) PDF, then (X_1, X_2) is the same as $(X, n - X)$. Now $X_1 + X_2 = n$ and $p_1 + p_2 = 1$. Using this new notation, we can rewrite Equation (8.15) as

$$Z^2 = \frac{(X_1 - np_1)^2}{np_1} + \frac{(X_2 - np_2)^2}{np_2}, \tag{8.17}$$

where we have used the fact shown in Equation (8.16) that

$$X_2 - np_2 = -(X_1 - np_1) = -(X - np_0) \implies (X_2 - np_2)^2 = (X - np_0)^2.$$

Example 8.4 If $n = 100$ and $p_0 = \frac{1}{2}$, then $np_0 = 50$ and $np_0(1 - p_0) = 25$. By Equations (8.13)–(8.17), we reject H_0 if

$$Z^2 = \frac{(X - 50)^2}{25} = \frac{(X_1 - 50)^2}{50} + \frac{(X_2 - 50)^2}{50} > \chi_1^2(.95) = 3.8415.$$

8.3.4 *m*-Category Multinomial Case

The extension to m categories for testing the null hypothesis

$$H_0 : \mathbf{p} = \mathbf{p}_0 = (p_{01}, p_{02}, \dots, p_{0m}) \tag{8.18}$$

results in a new test statistic, which we denote by $\mathcal{X}_m^2$:

$$\mathcal{X}_m^2 = \sum_{k=1}^{m} \frac{(X_k - np_{0k})^2}{np_{0k}} \sim \chi_{m-1}^2. \tag{8.19}$$

Note that the number of degrees of freedom (df, i.e. the parameter p of the $\chi^2(p)$ statistic) is one less than the number of categories m. Earlier we saw that with two categories, the χ^2 test statistic had df parameter $p = 1$.

Example 8.5 The number of murders in Houston each year from 2014 to 2017 was 242, 297, 303, and 269, respectively. Test the null hypothesis that $p_{0k} = \frac{1}{4}$ for $k = 1, 2, 3, 4$, i.e. the uniform distribution. Since $n = 1{,}111$, $np_{0k} = 277.75$. Hence,

$$\mathcal{X}_4^2 = \sum_{k=1}^{4} \frac{(X_k - 277.75)^2}{277.75} = 8.506 > \chi_3^2(0.95) = 7.815 \ ;$$

therefore, we reject the null hypothesis. The murder rate does not seem to have been constant over these four years. Perhaps a (local) quadratic trend might be considered, where the rate first increases and then decreases.

8.3.5 Goodness-of-Fit Test for a Parametric Model

Following Section 8.3.2, we used the computed bin probabilities of the parametric model $p(x|\theta)$ or $f(x|\theta)$ to compute the expected bin count as np_{0k}. We adopt the more conventional notation (o_k, e_k) in place of (X_k, np_{0k}); hence, our test statistic in Equation (8.19) is now

$$\mathcal{X}_m^2 = \sum_{k=1}^{m} \frac{(o_k - e_k)^2}{e_k}. \tag{8.20}$$

We consider two cases. In the first case, the null hypothesis completely specifies the probability vector $\mathbf{p}_0$ in Equation (8.18). In this case, the degrees of freedom $p = m - 1$, as in Example 8.5 above. In the second case, we must estimate $\mathbf{p}_0$ using MLE. This results in the estimates $\hat{p}_{0k} = o_k/n$. The degrees of freedom p will be either $m - 2$ or $m - 3$, depending upon whether the number of parameters in the PMF or PDF vector θ is 1 or 2. Note that the degrees of freedom p is reduced by 1 or 2 from the formula in the first case.

Example: For this example, the oracle knows we generated 50 standard normal data in **R** using `set.seed(123); x=rnorm(50)`. The nine bins extending from -2.0 to 2.5 with bin width of 0.5 give bin counts $(2, 6, 6, 12, 9, 8, 3, 3, 1)$. It is generally accepted that the CLT requires $e_k \geq 5$; hence, we collapsed the nine bins into the six intervals

$$(-\infty, -1), (-1, -0.5), (-0.5, 0), (0, 0.5), (0.5, 1), (1, \infty).$$

Since the entire real line $(-\infty, \infty)$ is the domain of the normal density, we extended the first and last bins to $\mp$ infinity, respectively.

There are two versions of the null hypothesis: $H_0 : X \sim N(0, 1)$ or $H_0 : X \sim N(\mu, \sigma^2)$. In the first case, the bin probabilities are computable, while in the second case, we replace the two unspecified parameters μ and σ^2 by their MLEs. In both cases, the alternative hypothesis is that the data are not well represented by the normal curve.

To compute the bin probabilities in the first case, set the bin edges in **R**

```
tk = c( -Inf, -1, -.5, 0, .5, 1, Inf)
pk = diff( pnorm( tk, 0, 1 ) )      (the bin probabilities).
```

In the second case, we need MLEs of μ and σ, which are $\bar{x} = 0.0344$ and $\hat{\sigma} = 0.9166$, respectively:

```
pk = diff( pnorm( tk, 0.0344, 0.9166 ) )
```

Thus, we have the counts, $\{o_k\}$, and expectations, $\{e_k = 50\,p_k\}$, given by Table 8.1. Thus, we compute in the two cases:

$$\mathcal{X}_6^2 = \sum_{k=1}^{6} \frac{(o_k - e_k)^2}{e_k} = 1.092 < 11.071 = \chi_{6-1}^2(0.95)$$

$$\mathcal{X}_6^2 = \sum_{k=1}^{6} \frac{(o_k - e_k)^2}{e_k} = 1.179 < 7.815 = \chi_{6-1-2}^2(0.95).$$

We fail to reject the null hypothesis in both cases. There is no reason not to work with the normal model for these data. Note the degrees of freedom were reduced from $m - 1 = 5$ by

Table 8.1 Data for testing the adequacy of a normal model.

Bin number	1	2	3	4	5	6
Bin count(o_k)	8	6	12	9	8	7
$N(0, 1)$ Bin probability	0.159	0.150	0.191	0.191	0.150	0.159
Bin expectation	7.933	7.494	9.573	9.573	7.494	7.933
$N(\bar{x}, s^2)$ Bin probability	0.130	0.150	0.205	0.209	0.160	0.146
Bin expectation	6.477	7.519	10.255	10.462	7.984	7.303

two more in the second case, since we estimated the two normal parameters by maximum likelihood.

8.3.6 Tests for Independence in Contingency Tables

Our final Pearson χ^2 test places each of n individuals or objects into one and only one "cell" of an $r \times s$ matrix, which is called a contingency table. The rows are characterized by the r sets $\{A_1, A_2, \ldots, A_r\}$, while the columns are characterized by the s sets $\{B_1, B_2, \ldots, B_s\}$. The number of individuals with characteristic A_i and B_j is denoted by o_{ij}. Note $\sum_{i=1}^{r} \sum_{j=1}^{s} o_{ij} = n$.

We wish to test whether the sets or characteristics A_i and B_j are independent of each other, that is,

$$H_0: \; p_{ij} = P(A_i \cap B_j) = P(A_i) \, P(B_j) \qquad \text{i.e. independent}$$
$$\text{versus } H_1: \; p_{ij} = P(A_i \cap B_j) \neq P(A_i) \, P(B_j) \qquad \text{for some } i, j.$$

Occasionally, the null hypothesis specifies all $r \times s$ probabilities p_{ij}, in which case $e_{ij} = n p_{ij}$ and the χ^2 test statistic, which is given by

$$\mathcal{X}_{r,s}^2 = \sum_{i=1}^{r} \sum_{j=1}^{s} \frac{(o_{ij} - e_{ij})^2}{e_{ij}}, \tag{8.21}$$

which has a χ^2 PDF with $rs - 1$ degrees of freedom.

Most often, we need to use MLE to estimate the unknown probabilities. We find that $\hat{p}_{ij} = o_{ij}/n$. Hence, $\widehat{P(A_i)} = \sum_{j=1}^{s} o_{ij}/n$ and $\widehat{P(B_j)} = \sum_{i=1}^{r} o_{ij}/n$. Therefore, we estimate the quantities e_{ij} by the formula

$$e_{ij} = n\hat{p}_{ij} = n \, \widehat{P(A_i)}\widehat{P(B_j)} = n \times \frac{o_{i\bullet}}{n} \times \frac{o_{\bullet j}}{n} = \frac{o_{i\bullet} \times o_{\bullet j}}{n}, \tag{8.22}$$

where we have adopted the "dot" notation to denote summation over that index. For example, $o_{i\bullet} = \sum_{j=1}^{s} o_{ij}$. We again use the formula in Equation (8.21) to compute the χ^2 test statistic. The degrees of freedom are now

$$df = (r-1)(s-1) = rs - r - s + 1,$$

since the sum of each row and column of the observed count matrix $\{o_{ij}\}$ and expected count matrix $\{e_{ij}\}$ exactly agree. Hence, we need only $r-1$ or $s-1$ of the row or column expectations to determine the remaining one. Because the sum of all the expectations equals n, we need to add 1 back to avoid double counting the redundant row constraint.

Example: Fisher (1936) re-analyzed a dataset collected by Tocher on 3883 Scottish children's hair color and sex, see Table 8.2.

Table 8.2 Contingency table for Scottish children.

Sex\hair color	Fair	Red	Medium	Dark	Jet black	Row total
Boys	592	119	849	504	36	2100
Girls	544	97	677	451	14	1783
Column total	1136	216	1526	955	50	3883

Table 8.3 Expected values assuming independence of sex and hair color.

Sex\hair color	Fair	Red	Medium	Dark	Jet black	Row total
Boys	614.37	116.82	825.29	516.48	27.04	2100
Girls	521.63	99.18	700.71	438.52	22.96	1783
Column total	1136	216	1526	955	50	3883

Following Equation (8.22), the matrix of expected values is given in Table 8.3. With these figures, the χ^2 statistic has $(2-1) \times (5-1) = 4$ degrees of freedom and is given by

$$\chi^2_{2,5} = \sum_{i=1}^{2} \sum_{j=1}^{5} \frac{(o_{ij} - e_{ij})^2}{e_{ij}}$$

$$= 10.467 > \chi^2_4(0.95) = 9.488 \; ;$$

therefore, we reject the null hypothesis. The p-value is `1-pchisq(10.467,4)` $=$ `0.033`. However, as a practical matter, with such a large sample size, the departure from independence is rather modest. There are slightly more boys with fair hair color and slightly more girls with medium and black jet hair colors than expected by independence. This may be an example of the notion that the null hypothesis is almost never exactly true; therefore, it can be expected to be rejected if n is very large.

8.4 Correlation Coefficient Tests and CIs

Given a random sample, $\{(X_1, Y_1), (X_2, Y_2), \ldots, (X_n, Y_n)\}$ from a bivariate normal PDF with correlation coefficient ρ, the maximum likelihood estimator of the sample correlation coefficient, R, is given by

$$R = \frac{\sum_{i=1}^{n}(X_i - \overline{X})(Y_i - \overline{Y})}{\sqrt{\sum_i(X_i - \overline{X})^2} \sqrt{\sum_i(Y_i - \overline{Y})^2}}. \tag{8.23}$$

In this section, we examine several tests and confidence intervals for the unknown correlation coefficient, ρ.

8.4.1 How to Test if the Correlation $\rho = 0$

The most common test performed is

$$H_0 : \rho = 0 \quad \text{versus} \quad H_1 : \rho \neq 0.$$

If the null hypothesis is rejected, not only are X and Y correlated, but they cannot be independent. Intuitively, the form of the critical region will be

$$C = \{(\mathbf{X}, \mathbf{Y}) : |R| > a \}. \tag{8.24}$$

Fisher provided a means of determining the threshold a by deriving the following approximate distribution result:

$$\frac{1}{2}\log\frac{1+R}{1-R} \approx N\left(\frac{1}{2}\log\frac{1+\rho}{1-\rho}, \frac{1}{n-3}\right). \tag{8.25}$$

Usually, the variance would also be a function of ρ. Here, it is not. This is another example of a *variance-stabilizing transformation*, which we saw first in Section 5.4.3 for the square root of a Poisson random variable.

Under the null hypothesis, the normal mean is 0, so C^c is

$$z_{\alpha/2} \le \frac{\frac{1}{2}\log\frac{1+R}{1-R} - 0}{\frac{1}{\sqrt{n-3}}} \le z_{1-\alpha/2},$$

which, when rearranged, gives the region (interval) for R itself as

$$\frac{e^{\frac{2z_{\alpha/2}}{\sqrt{n-3}}} - 1}{e^{\frac{2z_{\alpha/2}}{\sqrt{n-3}}} + 1} \le R \le \frac{e^{\frac{2z_{1-\alpha/2}}{\sqrt{n-3}}} - 1}{e^{\frac{2z_{1-\alpha/2}}{\sqrt{n-3}}} + 1} ;$$

hence, the critical region at the α-level for testing $\rho = 0$ is given by

$$C = \left\{ (\mathbf{X}, \mathbf{Y}) : |R| > \frac{e^{\frac{2z_{1-\alpha/2}}{\sqrt{n-3}}} - 1}{e^{\frac{2z_{1-\alpha/2}}{\sqrt{n-3}}} + 1} \right\}. \tag{8.26}$$

This critical region may be approximated by $|R| > z_{1-\alpha/2}/\sqrt{n-3}$. We can also use Equation (8.26) to solve for the sample size required to detect a non-zero value of ρ (i.e. reject $H_0 : \rho = 0$) as a function of n; see Figure 8.3.

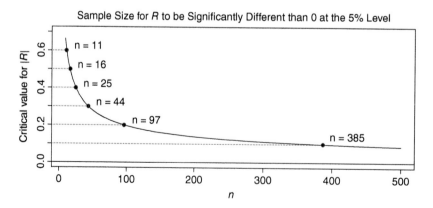

Sample Size for R to be Significantly Different than 0 at the 5% Level

Figure 8.3 Critical values for R as the sample size increases.

8.4.2 Confidence Intervals and Tests for a General Correlation Coefficient

To obtain a $100(1 - \alpha)\%$ confidence interval for ρ, we use Fisher's transformation in Equation (8.25)

$$z_{\alpha/2} \leq \frac{\frac{1}{2}\log\frac{1+R}{1-R} - \frac{1}{2}\log\frac{1+\rho}{1-\rho}}{\frac{1}{\sqrt{n-3}}} \leq z_{1-\alpha/2}, \tag{8.27}$$

which, when carefully rearranged, gives the $100(1 - \alpha)\%$ CI for ρ as the random interval

$$\frac{e^{\frac{-2z_{1-\alpha/2}}{\sqrt{n-3}}} - \frac{1-R}{1+R}}{e^{\frac{-2z_{1-\alpha/2}}{\sqrt{n-3}}} + \frac{1-R}{1+R}} \leq \rho \leq \frac{e^{\frac{-2z_{\alpha/2}}{\sqrt{n-3}}} - \frac{1-R}{1+R}}{e^{\frac{-2z_{\alpha/2}}{\sqrt{n-3}}} + \frac{1-R}{1+R}}. \tag{8.28}$$

To test the general hypotheses for some $-1 < \rho_0 < 1$,

$$H_0 : \rho = \rho_0 \quad \text{versus} \quad H_1 : \rho \neq \rho_0,$$

we would evaluate the quantity in Equation (8.27) at $R = r$ and $\rho = \rho_0$. If the the two inequalities hold, we would fail to reject the null hypothesis; otherwise, we would reject the null hypothesis. Equivalently, we check if ρ_0 is or is not in the confidence interval in Equation (8.28).

Examples: If $r = 0.3$ and $n = 25$, then the 95% confidence interval is $-0.108 \leq \rho \leq 0.622$. Thus we would fail to reject the null hypothesis $\rho = 0$, or any other value of ρ in the confidence interval. If $r = 0.75$ and $n = 100$, then the 95% confidence interval is $0.649 \leq \rho \leq 0.825$. We would clearly reject the null hypothesis $\rho = 0$. Our final example verifies the sample size given in Figure 8.3. If $r = 0.1$ and $n = 385$, then the 95% confidence interval is $0.0001 \leq \rho \leq 0.1980$; hence, we would reject the null hypothesis that $\rho = 0$.

8.5 Linear Regression

We will consider the case where the scientist chooses the set of experimental points $\{x_i, i = 1, \dots, n\}$ at which to measure the responses $\{Y_i, i = 1, \dots, n\}$. The results are quite similar to assuming a random sample from a bivariate normal PDF.

8.5.1 Least Squares Regression

Our data are the n pairs (x_i, Y_i), where x_i is not random. The linear model fits the data with the straight line equation

$$m(x) = a + b(x - \bar{x}) \quad \text{the straight-line model, with data}$$

$$Y_i | x_i = a + b(x_i - \bar{x}) + \epsilon_i, \quad \text{where } \epsilon_i \sim N(0, \sigma_\epsilon^2). \tag{8.29}$$

Note where $\bar{x}$ is inserted in these equations. The log-likelihood is given by

$$\ell(a, b, \sigma_\epsilon^2 | \epsilon) = -\frac{n}{2}\log 2\pi - \frac{n}{2}\log \sigma_\epsilon^2 - \frac{1}{2\sigma_\epsilon^2}\sum_{i=1}^{n} \epsilon_i^2, \tag{8.30}$$

where we assume the residuals $\{\epsilon_i\}$ are i.i.d. As before, the MLE for σ_ϵ^2 is

$$\tilde{\sigma}_\epsilon^2 = \frac{1}{n} \sum_{i=1}^n \epsilon_i^2. \tag{8.31}$$

Given $\tilde{\sigma}_\epsilon^2$, the MLEs for the intercept and slope maximize the final term in the log-likelihood in Equation (8.30), $-\sum \epsilon_i^2$, or equivalently,

$$(\hat{a}, \hat{b}) = \arg\min_{(a,b)} \sum_{i=1}^n \epsilon_i^2 = \arg\min_{(a,b)} \sum_{i=1}^n [Y_i - a - b(x_i - \bar{x})]^2. \tag{8.32}$$

The intercept, a, is determined by finding the root of

$$0 = \frac{\partial \sum \epsilon_i^2}{\partial a} = -2 \sum_{i=1}^n [Y_i - a - b(x_i - \bar{x})] = -2(n\bar{Y} - na - 0) \quad \text{or}$$

$$\boxed{\hat{a} = \bar{Y} \qquad \text{the MLE of the intercept},} \tag{8.33}$$

since $\sum(x_i - \bar{x}) = 0$.

Likewise, the slope is determined by finding the root of

$$0 = \frac{\partial \sum \epsilon_i^2}{\partial b} = -2 \sum_{i=1}^n (x_i - \bar{x})[Y_i - a - b(x_i - \bar{x})] \quad \text{or}$$

$$\boxed{\hat{b} = \frac{\sum_{i=1}^n (x_i - \bar{x})(Y_i - \bar{Y})}{\sum_{i=1}^n (x_i - \bar{x})^2} \qquad \text{the MLE of the slope},} \tag{8.34}$$

substituting $\bar{Y}$ for a.

Clearly, $\hat{a} = \bar{Y}$ is normal, and so is $\hat{b}$, since it is a linear combination of $Y_i - \bar{Y}$ with weights $\{w_i\}$, where

$$w_i = \frac{x_i - \bar{x}}{\sum_{j=1}^n (x_j - \bar{x})^2} \qquad \Rightarrow \qquad \hat{b} = \sum_{i=1}^n w_i (Y_i - \bar{Y}).$$

It is easy to check that the weights $\{w_i\}$ satisfy three formulae:

$$\sum_{i=1}^n w_i = 0 ; \quad \sum_{i=1}^n (x_i - \bar{x})w_i = 1 ; \quad \sum_{i=1}^n w_i^2 = \frac{1}{\sum_{j=1}^n (x_j - \bar{x})^2}. \tag{8.35}$$

Finally, we note from Equation (8.32) that the MLE approach is also called *least squares* (LS), since the criterion minimizes the sum of the squared residuals. The LS criterion is often applied without regards to any normality assumption.

8.5.2 Distribution of the Least-Squares Parameters

In order to conduct inference with the linear regression model, we need to understand the statistics of our three parameter estimators. We begin with the intercept, $\hat{a} = \bar{Y}$. From Equation (8.29), we have (noting $\sum(x_i - \bar{x}) = 0$)

$$\sum_{i=1}^n Y_i = na + b \sum_{i=1}^n (x_i - \bar{x}) + \sum_{i=1}^n \epsilon_i \quad \Rightarrow \quad \bar{Y} = a + \bar{\epsilon} \tag{8.36}$$

$$\hat{a} \sim N\left(a, \frac{\sigma_\epsilon^2}{n}\right);$$

(8.37)

hence, $\hat{a}$ is an unbiased estimator of the intercept a.

Next we consider the slope. Using Equations (8.36) and (8.35),

$$Y_i - \overline{Y} = [a + b(x_i - \overline{x}) + \epsilon_i] - [a + \overline{\epsilon}] = (\epsilon_i - \overline{\epsilon}) + b(x_i - \overline{x})$$

$$\hat{b} = \sum_{i=1}^{n} w_i(Y_i - \overline{Y}) = \sum_{i=1}^{n} w_i(\epsilon_i - \overline{\epsilon}) + b \sum_{i=1}^{n} w_i(x_i - \overline{x})$$

$$= \sum_{i=1}^{n} w_i\epsilon_i - \overline{\epsilon} \sum_{i=1}^{n} w_i + b = \sum_{i=1}^{n} w_i\epsilon_i + b$$

$$\hat{b} \sim N\left(b, \frac{\sigma_\epsilon^2}{\sum_i (x_i - \overline{x})^2}\right),$$

(8.38)

since $\sum w_i = 0$ and var $\sum_i w_i \epsilon_i = \sigma_\epsilon^2 \sum_i w_i^2$. Hence, $\hat{b}$ is an unbiased estimator of the slope.

Interestingly, $\hat{a}$ and $\hat{b}$ are uncorrelated, hence, independent. This follows directly since

$$\text{cov}(\hat{a}, \hat{b}) = \text{cov}\left(a + \overline{\epsilon}, b + \sum w_i\epsilon_i\right) = \text{cov}\left(\overline{\epsilon}, \sum w_i\epsilon_i\right) = E\left(\overline{\epsilon} \cdot \sum w_i\epsilon_i\right)$$

$$= E\left(n^{-1} \sum_j \epsilon_j \cdot \sum_i w_i\epsilon_i\right) = n^{-1}E\left(\sum_i w_i\epsilon_i^2 + \sum_{i \neq j} w_i\epsilon_i\epsilon_j\right)$$

$$= n^{-1}\sigma_\epsilon^2 \sum_i w_i + n^{-1} \sum_{i \neq j} w_i E[\epsilon_i\epsilon_j] = 0,$$

since $\sum_i w_i = 0$ and the ϵ_i are a random sample with mean 0.

Finally, we modify the MLE for σ_ϵ^2 in Equation (8.31) to obtain an unbiased estimator for residual variance, σ_ϵ^2, namely,

$$\hat{\sigma}_\epsilon^2 = \frac{1}{n-2} \sum_{i=1}^{n} [Y_i - \hat{a} - \hat{b}(x_i - \overline{x})]^2.$$

(8.39)

In an advanced course, we can prove this estimator of σ_ϵ^2 is independent of $\hat{a}$ and $\hat{b}$ and follows a χ^2_{n-2} distribution when normalized.

8.5.3 A Confidence Interval for the Slope

Using all the results of the previous section, we standardize the slope estimate $\hat{b}$ using the unbiased estimator of σ_ϵ^2 and obtain

$$T_{n-2} = \frac{(\hat{b} - b)/\sqrt{\sigma_\epsilon^2 / \sum (x_i - \overline{x})^2}}{\sqrt{\frac{n-2}{\sigma_{2\epsilon}}\hat{\sigma}_\epsilon^2/(n-2)}} = \frac{\hat{b} - b}{\sqrt{\hat{\sigma}_\epsilon^2 / \sum (x_i - \overline{x})^2}}.$$

(8.40)

Let $t = T_{n-2}(1 - \alpha/2)$. Then a $100(1 - \alpha)\%$ confidence interval is

$$\hat{b} - t\sqrt{\hat{\sigma}_\epsilon^2 / \sum (x_i - \overline{x})^2} < b < \hat{b} + t\sqrt{\hat{\sigma}_\epsilon^2 / \sum (x_i - \overline{x})^2}.$$

(8.41)

8.5.4   A Two-side Hypothesis Test for the Slope

To test $H_0 : b = b_0$ versus $H_1 : b \neq b_0$, we compute

$$T_{n-2} = \frac{\hat{b} - b_0}{\sqrt{\hat{\sigma}_\epsilon^2 / \sum (x_i - \bar{x})^2}}. \tag{8.42}$$

and observe if T_{n-2} falls in the interval $(-t, t)$ or not.

8.5.5   Predictions at a New Value

The predicted value (conditional mean) of Y at a new value x is given by

$$\hat{m}(x) = \hat{Y}|x = \hat{a} + \hat{b}(x - \bar{x})$$
$$E[\hat{m}(x)] = a + b(x - \bar{x}),$$

so it is unbiased for the true value $m(x) = a + b(x - \bar{x})$. The variance of the estimate is easy to compute since $\hat{a}$ and $\hat{b}$ are uncorrelated:

$$\text{var } \hat{m}(x) = \text{var } \hat{a} + (x - \bar{x})^2 \text{var } \hat{b} + 2(x - \bar{x})\text{cov}(\hat{a}, \hat{b})$$
$$= \frac{\sigma_\epsilon^2}{n} + (x - \bar{x})^2 \frac{\sigma_\epsilon^2}{\sum (x_i - \bar{x})^2} + 0 = \sigma_\epsilon^2 \left(\frac{1}{n} + \frac{(x - \bar{x})^2}{\sum (x_i - \bar{x})^2} \right).$$

Following Equation (8.38)

$$\hat{m}(x) \sim N \left(m(x), \sigma_\epsilon^2 \left(\frac{1}{n} + \frac{(x - \bar{x})^2}{\sum (x_i - \bar{x})^2} \right) \right). \tag{8.43}$$

We may use this result to find confidence intervals and test statistics in the same manner as we did for $\hat{b}$ in the previous two sections. Only the variance expression is changed:

$$T_{n-2} = \frac{\hat{m}(x) - m(x)}{\sqrt{\hat{\sigma}_\epsilon^2 \left(\frac{1}{n} + \frac{(x-\bar{x})^2}{\sum (x_i - \bar{x})^2} \right)}}. \tag{8.44}$$

8.5.6   Population Interval at a New Value

Complementing the previous section, we can model a **new measurement** at $x = x_{n+1}$ using our parameter estimates fit based upon the first n data pairs as

$$Y_{n+1}|x_{n+1} = a + b(x_{n+1} - \bar{x}) + \epsilon_{n+1},$$

which are all independent of $\hat{a}$, $\hat{b}$, and $\hat{\sigma}_\epsilon^2$. The estimated mean of Y_{n+1} is

$$\hat{Y}_{n+1}|x_{n+1} = \hat{a} + \hat{b}(x_{n+1} - \bar{x}).$$

To find the variance, we examine the zero-mean quantity

$$\hat{Y}_{n+1} - Y_{n+1} = (\hat{a} - a) + (\hat{b} - b)(x_{n+1} - \bar{x}) - \epsilon_{n+1} ;$$
$$E[\hat{Y}_{n+1} - Y_{n+1}]^2 = \text{var}(\hat{a}) + \text{var}(\hat{b})(x_{n+1} - \bar{x})^2 + \text{var}(\epsilon_{n+1}),$$

since all the covariance terms are zero by independence. Therefore,

$$\text{var}(\hat{Y}_{n+1} - Y_{n+1}) = \frac{\sigma_\epsilon^2}{n} + \frac{(x_{n+1} - \bar{x})^2 \sigma_\epsilon^2}{\sum (x_i - \bar{x})^2} + \sigma_\epsilon^2,$$

As with Equation (8.44), the relevant statistic for testing and forming a confidence interval is

$$T_{n-2} = \frac{Y_{n+1} - \hat{Y}_{n+1}}{\sqrt{\hat{\sigma}_\epsilon^2 \left(1 + \frac{1}{n} + \frac{(x_{n+1} - \bar{x})^2}{\sum (x_i - \bar{x})^2}\right)}}. \tag{8.45}$$

8.6 Analysis of Variance

Despite the name of this technique, the analysis of variance (ANOVA) provides an exact test of the equality of K normal means without having to use the Bonferroni inequality to maintain the type I error. The model and data collected take the form:

$$\{X_{ik}, \ i = 1, 2, \ldots, n_k; \ k = 1, 2, \ldots, K\}$$
$$X_{ik} \sim N(\mu_k, \sigma^2),$$

where $\sum_k n_k = n$. The hypotheses take the form

$$H_0 : \ \mu_k = \mu_0, \ k=1,\ldots,K \quad \text{versus}$$
$$H_1 : \ \text{not all } \mu_k \text{ are equal}.$$

The intuitive idea behind ANOVA is quite clever. Under the null hypothesis, all n data points X_{ik} have the same distribution $N(\mu_0, \sigma^2)$. Also under the null hypothesis, the K group means $\bar{X}_k$ are distributed $N(\mu_0, \sigma^2/n_k)$. Thus two independent estimates of σ^2 can be obtained using either the raw data or the group means. However, if the alternative hypothesis is true, then the variance estimate obtained from the group means will be inflated, since the true means are not equal. We make these ideas firm below.

The ANOVA is based upon the following simple algebraic identity, which gives a decomposition of the **total sum of squares** into the sum of the **within sum of squares** and the **between sum of squares**, that is, $T = W + B$. Let $\bar{X}$ be the grand sample mean, and $\bar{X}_k$ be the mean of the kth group of data. Then,

$$T = \sum_{k=1}^{K} \sum_{i=1}^{n_k} (X_{ik} - \bar{X})^2$$

$$= \sum_{k=1}^{K} \sum_{i=1}^{n_k} [(X_{ik} - \bar{X}_k) + (\bar{X}_k - \bar{X})]^2$$

$$= \sum_{k=1}^{K} \sum_{i=1}^{n_k} [(X_{ik} - \bar{X}_k)^2 + 2(X_{ik} - \bar{X}_k)(\bar{X}_k - \bar{X}) + (\bar{X}_k - \bar{X})^2]$$

$$= \sum_{k=1}^{K} \sum_{i=1}^{n_k} (X_{ik} - \bar{X}_k)^2 + 0 + \sum_{k=1}^{K} n_k (\bar{X}_k - \bar{X})^2$$

$$= \sum_{k=1}^{K} (n_k - 1)\, S_k^2 + \sum_{k=1}^{K} n_k\, (\overline{X}_k - \overline{X})^2 = W + B, \quad \text{where}$$

$$W = \sum_{k=1}^{K} (n_k - 1)\, S_k^2$$

$$B = \sum_{k=1}^{K} n_k\, (\overline{X}_k - \overline{X})^2,$$

since $\sum_i (X_{ik} - \overline{X}_k) = 0$ in the third line for all k. Here, S_k^2 is the unbiased estimator of σ^2 using only the data from the kth group.

Under the null hypothesis, the random variables, T, W, and B, all have χ^2 distributions. For example,

$$\frac{T}{\sigma^2} \sim \chi_{n-1}^2$$

$$\frac{W}{\sigma^2} = \sum_{k=1}^{K} \frac{(n_k - 1) S_k^2}{\sigma^2} = \sum_{k=1}^{K} \chi_{n_k-1}^2 \sim \chi_{n-K}^2,$$

since $\sum_k (n_k - 1) = n - K$. We claim W and B are independent, so by the MGF technique,

$$\frac{B}{\sigma^2} = \frac{T}{\sigma^2} - \frac{W}{\sigma^2}$$

$$\sim \chi_{n-1}^2 - \chi_{n-K}^2$$

$$\sim \chi_{K-1}^2.$$

Thus the test is conducted by using the $F_{r,s}$ random variable

$$F_{K-1,n-K} = \frac{\frac{B/\sigma^2}{K-1}}{\frac{W/\sigma^2}{n-K}} = \frac{B/(K-1)}{W/(n-K)} \tag{8.46}$$

and rejecting the null hypothesis at the α level when

$$\frac{B/(K-1)}{W/(n-K)} > f_{K-1,n-K}(1 - \alpha). \tag{8.47}$$

Example: Suppose four calculus classes with 25 students each at a community college use the same materials and examinations; however, one class uses graphing calculators. Is there an effect?

We first check the precision of the F statistic where we assume the 100 grades have been standardized; then $K = 4$, $n_k = 25$, and under the null hypothesis $X_{ik} \sim N(0, 1)$. Since $\sigma^2 = 1$, $W/(n - K)$ is also approximately equal to 1. In Figure 8.4, we see the result of 10,000 simulations, which fit the $F_{3,99}$ PDF very well. If the graphing calculators have an effect, how much larger must the class mean be to be detected? As a rough estimate, suppose $\overline{x}_1$ is 1.86 greater in the class using graphing calculators than the other three classes, which have the same centered mean of 0. Then we may check $B/24 = 2.70$, so the F statistic in Equation (8.47) is just greater than 2.696, which is the 95th percentile of the $F_{3,99}$ PDF. So a

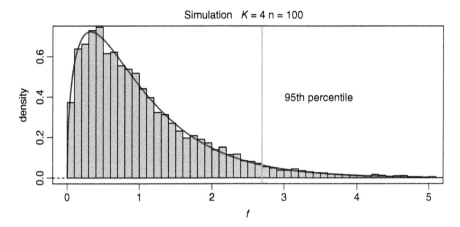

Simulation $K = 4$ $n = 100$

Figure 8.4 Ten thousand simulations of the F-statistic (8.46) with $K = 4$, $n_k = 25$, and $n = 100$ under the null hypothesis. $H_0 : \mu_k = \mu_0$, $k = 1, 2, 3, 4$.

relatively small improvement in the average performance of the 25 students in group 1 can be detected by the F-test, with correct overall type I error.

Problems

8.1 Find confidence intervals for paired and two-sample T-tests.

8.2 Show that the inequality in Equation (8.6) may be written in the form

$$\text{Prob}\left(\frac{\sqrt{n}\,\overline{X}}{\sqrt{\frac{(n-1)\,S^2}{n-1}}} < t_{0.975, n-1} \right) = 20\%,$$

where $\sqrt{n}\,\overline{X} \sim N(\sqrt{n}\mu_1, 1)$. The quantity follows the non-central T distribution, where the numerator is $N(\mu, 1)$ rather than $N(0, 1)$. For the parameters in the example,

```
pt( qt(.975,15), 15, 2.997865 ) = 0.20000;
```

where the non-centrality parameter $\mu = 2.997865$ was found by interpolation. Hence, $4\mu_1 = 2.997865$ or $\mu_1 = 0.7495$.

9

Topics in Statistics

We have now completed the core material for a one-semester course. The material in this final chapter serves a number of purposes. The first section provides a case study of a research project, namely, how to choose the bin width of a histogram in an optimal fashion. Other sections provide a selection of advanced and specialized topics, for which entire courses are available. An intuitive understanding will be valuable at this time.

9.1 MSE and Histogram Bin Width Selection

For most of our models, $f(x|\theta)$, we have been able to construct estimators $\hat{\theta}(\mathbf{x})$ that are unbiased, that is $E\hat{\theta}(\mathbf{x}) = \theta$. In practice, unbiased estimators are not common. In most complicated models, maximum likelihood estimators are only asymptotically unbiased. If $E\hat{\theta}(\mathbf{x}) = \theta$ is not unbiased, then the bias is defined to be

$$\text{Bias}(\hat{\theta}(\mathbf{x})) = E\hat{\theta}(\mathbf{x}) - \theta.$$

In this section, we generalize the variance criterion to the mean squared error (MSE). Then we illustrate its use on the problem of constructing an optimal histogram. We devote an unusually large amount of space to this problem, as it illustrates the research process, yet requires no more statistical theory than we have covered in this textbook.

9.1.1 MSE Criterion for Biased Estimators

For a single-parameter model, $f(x|\theta)$, the MSE criterion for evaluating the quality of $\hat{\theta}$ is defined as

$$\begin{aligned}
\text{MSE}(\hat{\theta}) &= E(\hat{\theta} - \theta)^2 \\
&= E[(\hat{\theta} - E\hat{\theta}) - (\theta - E\hat{\theta})]^2 \\
&= E(\hat{\theta} - E\hat{\theta})^2 - 2E[(\hat{\theta} - E\hat{\theta})(\theta - E\hat{\theta})] + E(\theta - E\hat{\theta})^2 \\
&= \text{var}\,\hat{\theta} - 2(\theta - E\hat{\theta})E(\hat{\theta} - E\hat{\theta}) + (\theta - E\hat{\theta})^2 \\
&= \text{var}\,\hat{\theta} - 2(\theta - E\hat{\theta}) \cdot 0 + (E\hat{\theta} - \theta)^2 \qquad \text{or}
\end{aligned}$$

$$\boxed{\text{MSE}(\hat{\theta}) = \text{var}\,\hat{\theta} + \text{Bias}(\hat{\theta})^2,} \tag{9.1}$$

Statistics: A Concise Mathematical Introduction for Students, Scientists, and Engineers, First Edition. David W. Scott.
© 2020 John Wiley & Sons Ltd. Published 2020 by John Wiley & Sons Ltd.

since $E(\hat\theta - E\hat\theta) = E\hat\theta - E\hat\theta = 0$. This MSE decomposition of error into variance and squared bias is universal. Of course, if $\hat\theta$ is in fact unbiased, then the MSE and the variance are equivalent.

Example: We compare the MSEs of three estimators of σ^2 with a $N(\mu, \sigma^2)$ random sample. Recall that $(n-1)S^2/\sigma^2 \sim \chi^2(n-1)$ has variance $2(n-1)$. Define

$$Y = cS^2, \qquad \text{for some constant } c; \text{ then}$$

$$EY = c\sigma^2 \implies \text{Bias}(c) = c\sigma^2 - \sigma^2$$

$$\text{Bias}(c)^2 = (c-1)^2\sigma^4.$$

Next, the variance is

$$\text{var}(Y) = c^2 \text{ var}(S^2) = c^2 \text{ var}\left(\frac{\sigma^2}{n-1} \cdot \overbrace{\frac{(n-1)S^2}{\sigma^2}}^{\chi^2(n-1)} \right)$$

$$= c^2 \cdot \frac{\sigma^4}{(n-1)^2} \cdot 2(n-1) = \frac{2c^2\sigma^4}{n-1}; \quad \text{hence;}$$

$$\boxed{\text{MSE}(c) = \frac{2c^2\sigma^4}{n-1} + (c-1)^2\sigma^4.}$$

A straightforward calculation shows that the MSE(c) is minimized when

$$c^* = \frac{n-1}{n+1};$$

hence, the MSEs for the three cases of interest are:

$$\hat\sigma_1^2 = \frac{1}{n-1}\sum (X_i - \bar{X})^2 \text{ is } \frac{2\sigma^4}{n-1} \quad \textbf{(unbiased estimator)}$$

$$\hat\sigma_2^2 = \frac{1}{n}\sum (X_i - \bar{X})^2 \text{ is } \frac{(2n-1)\sigma^4}{n^2} \quad \textbf{(MLE)}$$

$$\hat\sigma_3^2 = \frac{1}{n+1}\sum (X_i - \bar{X})^2 \text{ is } \frac{2\sigma^4}{n+1} \quad \textbf{(best } c^*\textbf{).}$$

We may check that, for these three estimators, the MSEs are in the order $\hat\sigma_3^2 < \hat\sigma_2^2 < \hat\sigma_1^2$. The unbiased estimator has the largest MSE.

9.1.2 Case Study: Optimal Histogram Bin Widths

Rosenblatt (1956) showed that probability histograms are consistent but biased estimators of the underlying density function. Scott (1979) showed how to estimate the MSE of a histogram, and then derived the "optimal" histogram, using only tools found in this textbook. Figure 9.1 displays the notation for a fixed-bin-width histogram with bins that begin at $t_0 = 0$.

We propose using the probability histogram,

$$\hat{f}_k(x) = \frac{v_k}{nh} \quad \text{for } x \in B_k,$$

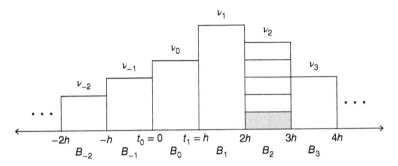

Figure 9.1 For a frequency histogram, the notation used to denote the locations of the bins $\{B_k\}$, the bin counts $\{v_k\}$, and the bin edges $\{t_k\}$.

as an estimator of the unknown density $f(x)$, for any value of $x \in \mathbb{R}^1$. We assume only that $f(x)$ has a continuous first derivative. In particular, we focus our attention on $\hat{f}(x)$ at a fixed point $x \in B_k = (kh, (k+1)h)$ as an estimator of $f(x)$ there. Since every data point, x_i, either is or is not in bin B_k, the bin count v_k is a binomial random variable with n "trials" (one for each of the n data points) and probability of "success" given by

$$p = p_k = \int_{B_k} f(s) \, ds = \int_{kh}^{(k+1)h} f(s) \, ds.$$

We are now in a position to compute the MSE of $\hat{f}(x)$. We begin by computing the variance, then the bias, for a fixed $x \in B_k$:

$$\text{var } \hat{f}(x) = \text{var } \frac{v_k}{nh} = \frac{1}{(nh)^2} \cdot np_k(1 - p_k) = \frac{p_k(1 - p_k)}{nh^2} \tag{9.2}$$

$$\text{Bias } \hat{f}(x) = E \frac{v_k}{nh} - f(x) = \frac{1}{nh} \cdot np_k - f(x) = \frac{p_k}{h} - f(x). \tag{9.3}$$

We are interested in how the choice of bin width, h, affects the entire error of the histogram over $-\infty < x < \infty$. To this end, we consider the integral of the MSE(x), namely, the integrated mean squared error (IMSE):

$$\text{IMSE} = \int_{-\infty}^{\infty} \text{MSE}(x) \, dx = \sum_{k=-\infty}^{\infty} \int_{B_k} \text{MSE}(x) \, dx$$

$$= \sum_{k=-\infty}^{\infty} \int_{B_k} \left(\text{var}(\hat{f}(x)) + \text{Bias}(\hat{f}(x))^2 \right) \, dx \equiv \text{IV} + \text{ISB}, \tag{9.4}$$

where we denote the integrated variance by IV and the integrated squared bias by ISB.

The integrated variance is easiest to obtain using the expression (9.2) for the var(x):

$$\text{IV} = \sum_{k=-\infty}^{\infty} \int_{B_k} \frac{p_k(1 - p_k)}{nh^2} \, dx = \sum_{k=-\infty}^{\infty} \frac{p_k(1 - p_k)}{nh^2} \times h$$

$$= \sum_{k=-\infty}^{\infty} \left[\frac{p_k}{nh} - \frac{p_k^2}{nh} \right] = \frac{1}{nh} \sum_{k=-\infty}^{\infty} p_k - \frac{1}{nh} \sum_{k=-\infty}^{\infty} p_k^2$$

$$\boxed{\text{IV} = \frac{1}{nh} - \frac{1}{n} \int_{-\infty}^{\infty} f(x)^2 \, dx + \cdots,} \tag{9.5}$$

since $\sum p_k = \int f(x)\,dx = 1$, and using the simple area approximation $p_k = hf(t_k)$ in the second sum to obtain the Riemann integral:

$$\frac{1}{nh}\sum_{k=-\infty}^{\infty} p_k^2 \approx \frac{1}{nh}\sum_{k=-\infty}^{\infty}(hf(t_k))^2 = \frac{1}{n}\sum_{k=-\infty}^{\infty}f(t_k)^2 \cdot h \xrightarrow{h\to 0} \frac{1}{n}\int_{-\infty}^{\infty}f(x)^2\,dx.$$

The ISB requires a bit more work. We focus on the MSE(x) for $x \in B_0$, and then generalize to other bins. We need a better approximation for p_0 than $hf(t_0)$. We begin with a Taylor series of $f(x)$ around $x = 0$ to obtain:

$$f(x) = f(0) + xf'(0) + \cdots$$

$$p_0 = \int_0^h f(x)\,dx = \int_0^h [f(0) + xf'(0) + \cdots]\,dx$$

$$= hf(0) + \frac{1}{2}h^2 f'(0) + \cdots.$$

Hence, the bias may be approximated by (ignoring higher order terms)

$$\text{Bias }\hat{f}(x) = \frac{p_0}{h} - f(x) \approx \frac{1}{h}\left[hf(0) + \frac{1}{2}h^2 f'(0)\right] - [f(0) + xf'(0)]$$

$$= \left(\frac{h}{2} - x\right)f'(0) + \cdots.$$

Now we compute the integrated squared bias for bin B_0:

$$\text{ISB}_0 = \int_0^h [\text{Bias }\hat{f}(x)]^2\,dx \approx \int_0^h \left(\frac{h}{2} - x\right)^2 f'(0)^2\,dx = \frac{1}{12}h^3 f'(0)^2.$$

For any other bin, B_k, the expression for ISB_k is identical, after replacing the bin boundary $0 = t_0$ with t_k in $f'(0)^2$. Therefore,

$$\boxed{\text{ISB} \approx \sum_{k=-\infty}^{\infty}\frac{1}{12}h^3 f'(t_k)^2 = \frac{1}{12}h^2\sum_{k=-\infty}^{\infty}f'(t_k)^2 \cdot h \xrightarrow{h\to 0} \frac{1}{12}h^2\int_{-\infty}^{\infty}f'(x)^2\,dx.}$$

Combining this result with the IV in Equation (9.5) gives

$$\boxed{\text{IMSE}(h, n) = \frac{1}{nh} + \frac{1}{12}h^2\int_{-\infty}^{\infty}f'(x)^2\,dx,} \qquad (9.6)$$

ignoring terms of higher order. For fixed n, the IMSE(h, n) is minimized when the bin width is given by

$$\boxed{h_n^* = \left[\frac{6}{n\int_{-\infty}^{\infty}f'(x)^2\,dx}\right]^{1/3}.} \qquad (9.7)$$

From Equations (9.7) and (9.6), we see that the optimal bin width decreases as the inverse of the cube root of the sample size n. For that bin width, the optimal IMSE decreases to 0 at the rate $O(n^{-2/3})$. This is a characteristic of non-parametric estimation, namely, that the rate of decrease of the error does not achieve the parametric rate of $O(n^{-1})$. The benefit is that the histogram "works" for any density without knowing its parametric form.

9.1.3 Examples with Normal Data

If the "unknown" density, $f(x)$, is in fact $N(\mu, \sigma^2)$, then we calculate that

$$\int_{-\infty}^{\infty} \left[\frac{\mathrm{d}\ \phi(x|\mu, \sigma^2)}{\mathrm{d}x} \right]^2 = \frac{1}{4\sqrt{\pi}\sigma^3} \ ; \qquad \text{hence,}$$

$$\boxed{\text{IMSE}(h, n) = \frac{1}{nh} + \frac{h^2}{48\sqrt{\pi}\sigma^3} \quad \text{and} \quad h_n^* \approx 3.5\ \sigma\ n^{-1/3}.} \tag{9.8}$$

Some examples of the IMSE(h, n) curves are displayed in Figure 9.2.

9.1.4 Normal Reference Rules for the Histogram Bin Width

John Tukey advocated assuming that the data are normal as a starting point for an analysis. In this section, we introduce three such procedures for the histogram that are all available as options in the **R** function `hist` through the argument `breaks`.

9.1.4.1 Scott's Rule
The only unknown in the theoretically optimal normal bin width reference rule in Equation (9.8) is the standard deviation σ. Scott (1979) proposed using the sample standard deviation, giving

$$\boxed{\hat{h}_n = 3.5\ \hat{\sigma}\ n^{-1/3} \qquad \textbf{Scott's rule.}} \tag{9.9}$$

9.1.4.2 Freedman–Diaconis Rule
For the $N(\mu, \sigma^2)$ density, the relationship between σ and the interquartile range (IQR) is

$$\text{IQR} = [\Phi(0.75) - \Phi(0.25)]\ \sigma = 1.3490\ \sigma \qquad \text{hence,}$$

$$h_n^* = 3.4908 \times \frac{\text{IQR}}{1.3490}\ n^{-1/3} = 2.588 \cdot \text{IQR} \cdot n^{-1/3}.$$

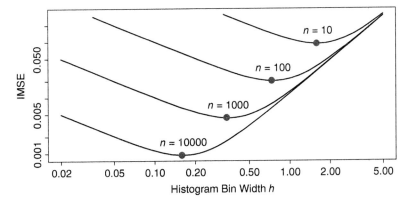

Figure 9.2 For several sample sizes and the $N(0, 1)$ density, the histogram IMSE curves as the bin width varies on a log-log scale. The red dots locate the best $h = h_n^*$.

Freedman and Diaconis (1981) chose to use the sample IQR rather than $\hat{\sigma}$ because it is robust to outliers. Furthermore, they chose to reduce the constant by 22.7%, giving

$$\boxed{\hat{h}_n = 2 \cdot \widehat{\text{IQR}} \cdot n^{-1/3}} \qquad \textbf{Freedman–Diaconis rule}. \qquad (9.10)$$

9.1.4.3 Sturges' Rule

The default rule for the **R** hist function is Sturges' rule; see Sturges (1926). Sturges used the binomial PMF with $p = \frac{1}{2}$ as a discrete approximation to a standard normal PDF, when rounded. The Binom(m, p) in this case is

$$p(x) = \binom{m}{x} p^x (1-p)^{m-x} = \binom{m}{x} \frac{1}{2^m}, \quad x = 0, 1, 2, \ldots, m ;$$

note that we use m in the binomial PMF rather than n, which we reserve for the sample size. As is clear from the middle column of Figure 3.6, this PMF looks normal because the combinatorial coefficients in the mth row of Pascal's triangle look normal. Sturges took those coefficients to be an "ideal" normal set of bin counts; hence, the corresponding total sample size equals

$$n = \binom{m}{0} + \binom{m}{1} + \binom{m}{2} + \cdots + + \binom{m}{m-1} + \binom{m}{m}$$

$$= \sum_{\ell=0}^{m} \binom{m}{\ell} = \sum_{\ell=0}^{m} \binom{m}{\ell} 1^\ell 1^{m-\ell} = (1+1)^m = 2^m \qquad (9.11)$$

by the binomial theorem. Since the mth row has $m + 1$ bins, we can solve for the number of bins, call it $k = m + 1$, using Equation (9.11) to obtain

$$n = 2^m = 2^{k-1} \qquad \boxed{k_n = 1 + \log_2(n)} \qquad \textbf{Sturges' rule}. \qquad (9.12)$$

There is nothing random in Sturges' rule. The number of bins grows very slowly as the sample size n increases.

Sturges' rule is for the number of bins. It may be converted to a bin width rule, for example, by using the sample range and computing

$$\hat{h}_n = \frac{x_{(n)} - x_{(1)}}{k_n} = \frac{x_{(n)} - x_{(1)}}{1 + \log_2(n)}.$$

Again, depending upon the behavior of the sample range, the bin width decreases much more slowly than in the previous two rules.

9.1.4.4 Comparison of the Three Rules

We wish to compare these three rules with normal data for a range of sample sizes

$$n = 2^7 = 128 \qquad n = 2^{14} = 16{,}384 \qquad n = 2^{21} = 2{,}097{,}152.$$

To avoid issues of the sample range of normal data, we instead use the three rules on the Beta$(5, 5)$ PDF, which is very close to a $N(1/2, 1/44)$ PDF. Since the support of a Beta PDF is always $(0, 1)$, we convert the Scott and Freedman–Diaconis bin-width rules into number-of-bins rules by taking the reciprocal of $\hat{h}$ and rounding up. For this density, the IQR $= 0.2161$.

Using the exact IMSE formula Equation (3.26) in Scott (2015), we display the IMSE as a function of the number of bins in Figure 9.3. We superimpose the selections made by the

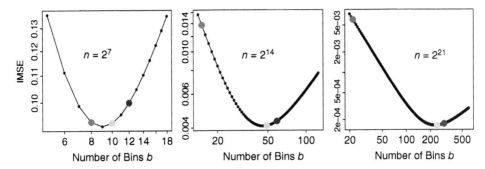

Figure 9.3 Comparison of the number of bins recommended by the Sturges (●), Scott (◐), and Freedman–Diaconis (●) rules for a Beta(5, 5) PDF plotted against the exact IMSE values.

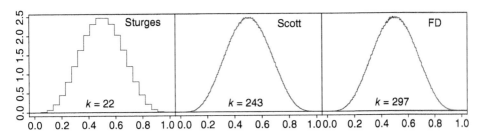

Figure 9.4 Three histograms of a Beta(5, 5) sample with $n = 2^{21}$.

three rules onto these curves. In each case, Sturges' rule selects the fewest bins, while the Freedman–Diaconis rule selects the most bins. Scott's rule is in the middle and is closest to the true minimizer in all three cases of varying sample sizes.

The most striking feature is that even with more than two million data points, Sturges' rule selects only 22 bins; see Figure 9.4. Intuitively, this choice is far from optimal and compresses the data into too few bins, squandering most of the information in the dataset. It is perhaps an unfortunate coincidence that for samples common in textbooks ($10 < n < 250$), the three rules select about the same number of bins. However, for sample sizes common in modern data science, Sturges' rule is to be avoided.

It turns out that the normal PDF is one of the easiest densities to estimate, since it is symmetric and has no unusual features, such as skewness or multimodality. The actual "optimal" histogram is likely to require more bins (that is, a smaller bin width) than Scott's rule in such situations. There are advanced algorithms, such as cross-validation, for finding such choices; see Scott (2015), Chapter 3. But for exploratory purposes or very large samples, the Scott and Freedman–Diaconis rules are excellent starting places.

9.2 An Optimal Stopping Time Problem

In this section, we consider a practical decision problem, namely, searching for the best option among n choices, but under certain order constraints. While we assume that the

n possibilities are presented to us in random order, the major constraint is that we must decide then and there whether to choose the current option or not. Furthermore, once we decline a choice, it is lost forever. We treat n as fixed and known; that is, we have enough resources to consider n options, but no more.

This problem has application in many aspects of life, although in the statistics and operations research literature, it is often set in the context of hiring. To make the model precise, we assume that we can perfectly rank the n choices (with no ties); however, we assume that we begin with no prior knowledge of the distribution of the quality of the n options. This model is conducive to analysis, but is obviously conservative. Whatever the solution, we should expect to be able to make a slightly better decision.

Let R_k denote the true rank of the kth option, which is unknown until we have seen all n possibilities. If we always choose the first option given to us, then the probability we have selected the "best" choice is only $1/n$.

At the kth step, we assume that we can insert the kth option into the previously ordered list. We denote the indices of the choices that we have seen as $\{i_1, i_2, \ldots, i_{k-1}\}$. After evaluating the kth option, we do not know R_{i_k} exactly, but rather its **relative rank**, which we denote by Y_k. If this choice is better than all the previous $k-1$ options, then $Y_k = 1$; otherwise, $Y_k \in \{2, 3, \ldots, k\}$.

The optimal strategy is to watch and patiently learn about the scope/range of available options; then, starting at the mth step, pick the first candidate that is better than all previous choices. Such a strategy will fail if the best choice was among the first $m-1$ candidates. In that case, $Y_k > 1$ for all $k = m, m+1, \ldots, n$. At that point, we can select the final candidate, but have failed at our goal. The strategy can also fail if $Y_k = 1$, i.e. the best candidate so far, but the absolute best choice is set to appear later.

What is the best choice for m? And for m^*, what is the probability that we have selected the absolute top candidate? We may compute this exactly by a conditional argument:

$$P(\text{select best}) = \sum_{k=m}^{n} P(\text{stop at } k | k \text{ is the best}) \, P(k \text{ is the best})$$

$$= \sum_{k=m}^{n} P(Y_\ell > 1, \ell = m, \ldots, k-1) \, P(R_{i_k} = 1)$$

$$= \sum_{k=m}^{n} \frac{m-1}{k-1} \times \frac{1}{n} \, ;$$

that is, while clearly $Y_k = 1$ since $R_{i_k} = 1$, in order for us to reach the kth possibility, the best of first $k-1$ choices observed so far must have been among the first $m-1$ choices. Now while the sum $\sum_k 1/k$ is (slowly) divergent, it is well-approximated by

$$\sum_{k=1}^{n} \frac{1}{k} = \log n + \gamma + O(n^{-1}),$$

where $\gamma \approx 0.5772$ is Euler's constant. Thus,

$$P(\text{select best}) = \frac{m-1}{n} \sum_{k=m}^{n} \frac{1}{k-1} = \frac{m-1}{n} \sum_{k=m-1}^{n-1} \frac{1}{k}$$

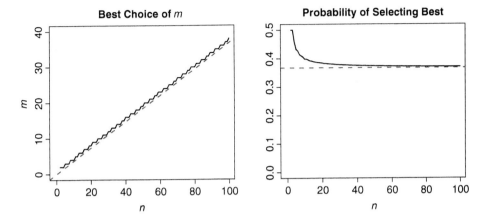

Figure 9.5 (Left) Optimal stopping point m as a function of the population size n compared to a straight line with slope $1/e$; (right) probability of selecting the best candidate using m^* compared to $p = 1/e$.

$$= \frac{m-1}{n}\left[\sum_{k=1}^{n-1}\frac{1}{k} - \sum_{k=1}^{m-2}\frac{1}{k}\right]$$

$$\approx \frac{m-1}{n}\log\left(\frac{n-1}{m-2}\right).$$

Now the function $(x/n)\log(x/n)$ is maximized when $x = n/e$, in which case the value of the function is $1/e = 36.8\%$; see Figure 9.5.

Thus the (surprising?) result is that we have a nearly constant probability of 36.8% of selecting the best choice, no matter how large n. If we, in fact, can reconsider earlier choices, we can improve our probability. Perhaps it is optimistic to believe we can perfectly rank choices. For example, if we are searching for a life partner over a period of several years, our ranking algorithm may not be constant over time.

9.3 Compound Random Variables

We saw in Section 2.2.5 how complicated probabilities can be computed by introducing a partition of the sample space. In this section, we discuss a related idea for expectation of random variables. We use the result to analyze the sum of random variables, where the number of r.v.s is itself random.

9.3.1 Computing Expectations with Conditioning

Suppose the random variables X and Y have joint PDF $f_{X,Y}(x,y)$, and we wish to compute EY. Since $f_{X,Y}(x,y) = f_{Y|X}(y|x)f_X(x)$,

$$EY = \int_x\int_y y\,f_{X,Y}(x,y)\,dy\,dx$$

$$= \int_x\left[\int_y y\,f_{Y|X}(y|x)\,dy\right]f_X(x)\,dx$$

$$= \int_x [E_{Y|X}(Y|X=x)] f_X(x)\, dx$$

$$= E_X\, E_{Y|X}(Y|X). \tag{9.13}$$

Carefully parse this expression and the use of the upper case random variables involved in the expectation. The same result holds if X and/or Y are discrete.

9.3.2 Sum of a Random Number of Random Variables

A compound random variable has a parameter that is itself a random variable. We limit our consideration to the sum of a random sample, where the number of terms, N, follows the PMF $p_N(n)$ and is independent of X_i:

$$\boxed{Y = \sum_{i=1}^{N} X_i.}$$

Its moments are

$$EY = E_N\, E_{Y|N}(Y|N)$$

$$= \sum_{n=0}^{\infty} E_{Y|N}(Y|N=n)\, p_N(n)$$

$$= \sum_{n=0}^{\infty} [n\, \mu_X]\, p_N(n)$$

$$\boxed{EY = \mu_N\, \mu_X.}$$

To find the variance of Y, we compute EY^2 in a similar fashion:

$$EY^2 = \sum_{n=0}^{\infty} E_X \left[\sum_{i=1}^{n} X_i^2 + \sum_{i \neq j} X_i X_j \right] p_N(n)$$

$$= \sum_{n=0}^{\infty} [n\,(\sigma_X^2 + \mu_X^2) + n(n-1)\mu_X^2]\, p_N(n)$$

$$= \mu_N\,(\sigma_X^2 + \mu_X^2) + (\sigma_N^2 + \mu_N^2 - \mu_N)\mu_X^2$$

$$= \mu_N\, \sigma_X^2 + \sigma_N^2\, \mu_X^2 + \mu_N^2\, \mu_X^2\, ; \qquad \text{hence,}$$

$$\text{var } Y = EY^2 - [EY]^2$$

$$\boxed{\text{var } Y = \mu_N\, \sigma_X^2 + \sigma_N^2\, \mu_X^2.}$$

Notice that if the PMF of N is concentrated on one value of n so $\sigma_N^2 = 0$, then the var $Y = n\sigma_X^2$ as usual. Otherwise, the variance is increased by the amount $\sigma_N^2\, \mu_X^2$.

Example: Compound Poisson distribution. Suppose $N \sim \text{Pois}(m)$. Then

$$\boxed{EY = m\, \mu_X}$$

$$\boxed{\text{var } Y = m\, \sigma_X^2 + m\, \mu_X^2.}$$

This model has utility when considering the risk assumed by an insurance company insuring a pool of individuals.

9.4 Simulation and the Bootstrap

We have made extensive use of simulation to understand a random process, to study its average behavior and variability, and to verify analytical results. These tasks all made use of an explicit model, $F_X(x|\theta)$, from which we generated many representative samples and summarized features of the sample.

Just as often, we wish to analyze a given dataset, $\{x_1, \dots, x_n\}$, without an explicit model. Sometimes the analysis does not lend itself to easy theoretical analysis. A natural candidate in lieu of a parametric model of the CDF is the empirical CDF, which is defined as

$$F_n(x) = \frac{\text{number of samples} \le x}{n}.$$

The empirical CDF looks like the discrete CDF in Figure 3.3, but with jumps of size $1/n$ at the n data values. Sampling from $F_n(x)$ is called **bootstrapping**. For a fixed value of x, it is easy to check that $F_n(x)$ is a binomial PMF with $p = F(x)$, since $Pr(X \le x) = F(x)$; hence, $F_n(x)$ is an unbiased estimator of $F(x)$.

The details of generating a bootstrap sample (or re-sample), denoted by

$$\tilde{\mathbf{x}} = (\tilde{x}_1, \tilde{x}_2 \dots, \tilde{x}_n)$$

are interesting. The empirical CDF places equal mass of $1/n$ on each value in the sample, $\mathbf{x} = \{x_1, x_2, \dots, x_n\}$. Hence, it is a discrete distribution. The bootstrap sample will take on values only in that list, some more than once, some not at all. In fact, if the bootstrap sample is also of length n, then the probability x_i is **not selected** at each step is $1 - 1/n$; hence, the probability it is not included in the bootstrap sample is

$$Pr(x_i \text{ is not in } \tilde{\mathbf{x}}) = \left(1 - \frac{1}{n}\right)^n \approx e^{-1} = 36.8\%.$$

Thus the bootstrap sample includes only about 63.2% of the original data values. We have often assumed that there are no ties in our random data, but this is not the case here. It is easy to generate a bootstrap sample of size n in **R** via

```
xB = x[ sample( n, n, replace=TRUE ) ],
```

as the **R** function `sample` returns a list of integers from 1 to n selected randomly, with replacement.

Example: Let us conduct a small experiment to illustrate some of the benefits and cautionary aspects of the bootstrap in practice. Suppose we are interested in summarizing a dataset through its mean and its robust cousin, the median. For a symmetric distribution, these are the same. For the normal distribution, we know $\overline{X}$ is better, but how much? We examine this question in two ways: first, with simulation, and second, based on a bootstrap analysis with a single data vector generated by `set.seed(132)`; `x=rnorm(101,5,1)`, for which $\overline{x} = 5.093$, $x_{0.5} = 5.001$, and $\hat{\sigma} = 0.909$. We selected 10^5 simulations and bootstrap re-samples.

In Figure 9.6(a), we see that the histogram matches the theoretical PDF for $\overline{X} \sim N(5, 1/101)$ shown in blue. However, in frame (b), the bootstrap histogram is shifted to $\overline{x}$ and narrower. This confirms our intuition that the bootstrap analysis will

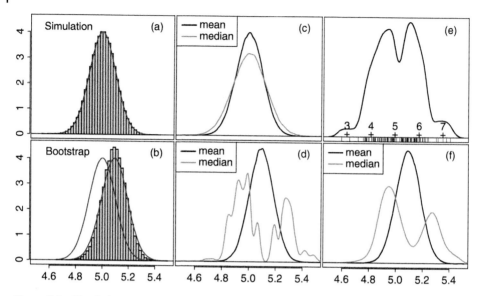

Figure 9.6 Simulation and bootstrap analysis of the sample mean and sample median for a $N(5, 1)$ PDF with $n = 101$ points; see text. Histograms and smoothed histograms are displayed.

mimic the features in the single data sample available. In frame (c), we see that the median has 57% ($\pi/2$ theoretically) greater variability than the sample mean. However, the picture in frame (d) for the bootstrap is less clear. Since the sample median of the bootstrap re-sample will always be the 51st point, it will always be one of the 101 values. (In fact, it took on only 25 values 99% of the time.) Furthermore, a typical bootstrap sample here has only 64 unique values; hence, the smoothed histogram will not be as smooth as in frame (c). If we examine the smoothed histogram of the bootstrap dataset in frame (e), we see that by chance the data are bimodal. The bootstrap clearly picks up that feature as it re-samples. The sample median tends to be slightly greater or less than the sample mean. One way to break ties is to add a little bit of normal noise to every bootstrap re-sample. The result is shown in frame (f). Of course, the bimodal feature of the sample median is not eliminated. This technique is called the *smoothed bootstrap*, and several choices of the normal blurring should be attempted.

The main uses of the bootstrap are to gauge the stability and reproducibility of an analysis. We make good use of these later in Figure 9.8.

9.5   Multiple Linear Regression

In a future statistics course, we will learn how to add more predictor variables to the linear regression model, which specifies a more complicated linear equation for the conditional mean. Examples of the multiple linear regression model with an intercept, β_0, and p predictors, $\{x_1, x_2, \ldots, x_p\}$, are

$$m(\mathbf{x}) = \beta_0 + \sum_{j=1}^{p} \beta_j x_j \tag{9.14}$$

$$m(\mathbf{x}) = \beta_0 + \beta_1 x_1 + \beta_2 x_2 + \beta_3 x_1^2 + \beta_4 x_2^2 + \beta_5 x_1 x_2. \tag{9.15}$$

The second example uses only two predictors but is considered to be *linear in the parameters*. It is a full quadratic model using the $p = 2$ variables, but includes two linear terms, two quadratic terms, and one interaction term. So $\mathbf{x}$ is coded as $(1, x_1, x_2, \ldots, x_p)^T$ for the model in Equation (9.14) and as $(1, x_1, x_2, x_1^2, x_2^2, x_1 x_2)^T$ for the model in Equation (9.15).

As before, the n design points $\{\mathbf{x}_1, \ldots, \mathbf{x}_n\}$ are taken as non-random. Given a sample of n points, $\{(\mathbf{x}_i, y_i), i = 1, 2, \ldots, n\}$, we use least-squares to fit the $p + 1$ β_j coefficients. If we put the data and model into matrix form,

$$\mathbf{Y} = \begin{pmatrix} y_1 \\ y_2 \\ \vdots \\ y_n \end{pmatrix} \qquad \mathbf{X} = \begin{pmatrix} 1 & x_{11} & x_{12} & \cdots & x_{1p} \\ 1 & x_{21} & x_{22} & \cdots & x_{2p} \\ \vdots & \vdots & \vdots & \vdots & \vdots \\ 1 & x_{n1} & x_{n2} & \cdots & x_{np} \end{pmatrix}, \qquad \text{then} \quad \mathbf{Y} = \mathbf{X}\,\beta + \epsilon,$$

where the n i.i.d. random variables $\epsilon_i \sim N(0, \sigma_\epsilon^2)$. The least-squares solution for β that minimizes $\epsilon^T \epsilon = \sum_i \epsilon_i^2$ is given by the matrix equation

$$\boxed{\hat{\beta} = (\mathbf{X}^T \mathbf{X})^{-1} \mathbf{X}^T \mathbf{Y}.}$$

The statistics of this estimator are beyond the scope of this course, but if we accept that $E[\,\epsilon\,] = \mathbf{0}_n$, then

$$E[\,\mathbf{Y}] = E[\,\mathbf{X}\,\beta + \epsilon]$$

$$= \mathbf{X}\,\beta + E[\,\epsilon\,] \quad \text{since the design matrix } \mathbf{X} \text{ is not random}$$

$$= \mathbf{X}\,\beta\,;$$

$$\text{hence,} \quad E[\,\hat{\beta}] = E[(\mathbf{X}^T \mathbf{X})^{-1} \mathbf{X}^T \mathbf{Y}\,]$$

$$= (\mathbf{X}^T \mathbf{X})^{-1} \mathbf{X}^T\, E[\,\mathbf{Y}\,] \qquad \text{by linearity}$$

$$= (\mathbf{X}^T \mathbf{X})^{-1} \mathbf{X}^T\, \mathbf{X}\,\beta$$

$$= \beta,$$

so the estimator is unbiased. Note we assume $\mathbf{X}$ is of full rank.

The **R** functions lsfit and lm can be used to perform these calculations. Software that performs tests such as $H_0: \beta[2:(p+1)] = \mathbf{0}_p$ and attempts to find a subset of the variables that are sufficient for prediction may be found in **R** functions such as step, which performs stepwise regression. (We seldom wish to test that the intercept β_1 is zero.) A small example is given in the next section.

9.6 Experimental Design

A special use of multiple regression is to determine optimal settings for manufacturing processes. The goal is find the best value of $\mathbf{x}$ that results in the highest average output. For a manufacturing process with two settings, the model given in Equation (9.15) will find that setting.

We analyze a subset of an example given in Devore and Farnum (1999). Their problem 10.2.13 examines the influence on "wet mold strength" of (1) the percentage of sand added $\{0\%, 15\%, 30\%\}$ as well as (2) the casting hardness setting. (A third variable measuring added carbon fiber is omitted.) The experiment was run $n = 18$ times with various settings. If we standardize both **Y** and the variables in **X**, then the **R** functions

```
ans = lsfit( scale(Y), scale(X[,-1]), intercept=TRUE )    or
ans = lm( scale(Y)~ scale(X[,-1]) )    returns
```

$$\hat{\beta} = (0.00, 3.41, 6.26, 0.80, -5.53, -5.07)^T.$$

Using this estimate, we computed the surface of average responses as $\mathbf{x}^T \hat{\beta}$, where **x** takes on values on a grid over the support of the data; see Figure 9.7. This surface is not the "ideal" elliptical shape with a single maximizer in the center. Instead, the best response seems to have no added sand; see the black curve in the right frame of Figure 9.7. However, when the added sand was 0%, all the experiments were run with the casting hardness between 61 and 69. Therefore, it is not clear if the predictions should be believed for larger values of casting hardness where the maximum appears to be headed. In fact, the three observed strengths at (0, 69) were (48, 55, 68), while the predicted average strength is 62.5. At the added sand value of 30%, a secondary maximum of 30.86 is observed; see the red line in the right frame of Figure 9.7.

On the other hand, none of the β coefficients are found to be significant, so that the design of this experiment can hardly be characterized as a success story. The JMP experimental design module includes dozens of success stories, as well as innovative algorithms for selecting values for n and $\mathbf{x}_i$; see JMP® (1989–2019).

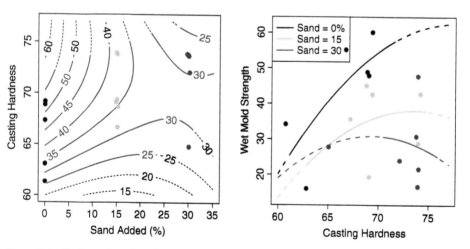

Figure 9.7 Surface of predicted average mold strength as a function of the two predictor variables. The contours and curves are shown as dotted lines where the standard deviation of the prediction exceeds 2.5.

9.7 Logistic Regression, Poisson Regression, and the Generalized Linear Model

Linear regression and its extension to multiple regression, which was discussed in Section 9.5, are powerful techniques for prediction. These methods rely mostly upon an assumption of normal errors or residuals. However, there are several cases where the responses have a special structure that requires other assumptions.

The data are the n pairs (x_i, Y_i) as before. First, we focus on predicting the **probability of an outcome** given the value of x. Clearly, $Y_i \in \{0, 1\}$ and the prediction should satisfy $E(\hat{Y}_i | X = x_i) \in [0, 1]$, a constraint which linear regression does not obey. Instead, we may consider a linear prediction of the log-odds, namely, $\log p/(1 - p)$, which is 0 when $p = 1/2$ and in $\mathbb{R}^1$ otherwise. Thus we have the model

$$\log \frac{p}{1 - p} = a + bx \quad \Rightarrow \quad p\,(1 + e^{a+bx}) = e^{a+bx} \quad \text{or}$$

$$\boxed{p = \frac{1}{1 + e^{-(a+bx)}}} \quad \text{Logistic regression.}$$

Another situation is illustrated in the final frame of Figure 1.5. Here, Y_i is a non-negative integer count of the number of O-ring failures, which we might model by a Poisson PMF. A linear prediction equation could model the log $E[Y|X = x]$, giving the so-called log-linear model equation

$$\boxed{E[Y|X = x] = e^{a+bx}} \quad \text{Poisson regression.}$$

Note the linear prediction in Figure 1.5 does not obey the non-negativity constraint.

The **R** function *glm*, which stands for **generalized linear model**, uses maximum likelihood to fit either of these models. The Bernoulli or Poisson PMFs are employed to compute the likelihood. For the space shuttle data, let x denote the temperature and y the number of O-rings that failed ($n = 23$). The glm fit for the Poisson regression is displayed in Figure 9.8

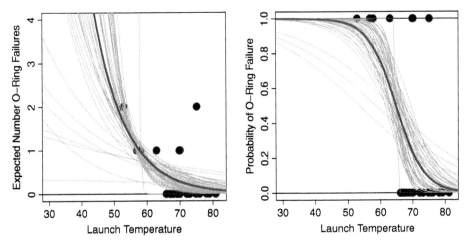

Figure 9.8 (Left) Poisson and (right) logistic regression models fitted by the **R** function `glm` to the space shuttle data (red line). Fifty bootstrap fits are superimposed (green lines).

and was generated by the commands `ans=glm( y~x,family=Poisson)`, with predictions `ypred=predict(ans,data.frame(x=30:82),type='response')`, and plotted by `plot(x,y,xlim=c(30,82)); lines(30:82,ypred)`.

For the logistic model, we modify the y vector to be only 0s and 1s. We accomplish this with the **R** command `y=ifelse(y==0,0,1)`. The logistic regression uses `family=binomial`. Both models strongly suggest serious O-ring problems at a freezing temperature. However, it is never a good idea to put too much stock in model extrapolations far beyond the data. Here, the 50 bootstrap fits displayed suggest the predictions are reliable.

9.8 Robustness

In practice, especially with the large datasets encountered in modern data science, bad data are a reality that must be accounted for. One or more outliers with univariate data can inflate both $\bar{x}$ and s^2. Recall that $\bar{x}$ minimizes the criterion

$$\bar{x} = \arg\min_c \sum_{i=1}^{n} (x_i - c)^2.$$

Since the residuals are squared, it is intuitive that outliers will have a disproportionate influence. Replacing the square with absolute value in the criterion decreases the influence of outliers, and gives a different estimator

$$x_{0.5} = \arg\min_c \sum_{i=1}^{n} |x_i - c|,$$

namely, the sample median. If the sample size n is odd, there is a clever proof of this fact. Let the criterion be defined by

$$g(c) = \sum_{i=1}^{n} |x_i - c| = \sum_{i:x_i<c} (c - x_i) + \sum_{i:x_i>c} (x_i - c) \text{ ; hence,}$$

$$\frac{\partial g(c)}{\partial c} = \sum_{i=1}^{n} I(x_i < c) - I(x_i > c),$$

which vanishes when the number of points greater than and less than c are both $(n-1)/2$. That means that $c^* = x_{0.5}$, as claimed. When n is even, the average of the two middle points is generally selected.

The median was in general use a century ago, but Fisher's proof of the optimality of MLE, and hence $\bar{x}$, turned the tide. For a random sample, we have

$$\bar{X} \approx N\left(\mu_X, \frac{\sigma_X^2}{n}\right) \quad \text{and} \quad X_{0.5} \approx N\left(\mu_X, \frac{1}{4f(\mu_X)^2\,n}\right),$$

as shown in more advanced textbooks; see Casella and Berger (2002). Thus for normal samples, in particular, the variance of the sample median is $\pi\sigma_X^2/2n$, which is larger by the fraction $\pi/2$ than the variance of $\bar{X}$, or 25.3% larger on the standard deviation scale. But of more importance, the median is unchanged if the largest sample goes off to $+\infty$. A more sophisticated analysis shows the median is robust to a large fraction of "bad" data, in fact, almost half.

For a robust estimator of the scale, there are two reasonable choices:

$$\text{MAD} = \text{median } |x_i - x_{0.5}| \qquad \textbf{median absolute deviation}$$

$$\text{IQR} = x_{0.75} - x_{0.25} \qquad \textbf{interquartile range}.$$

For a normal sample, these may be scaled to provide robust estimators for the standard deviation σ_x via IQR/1.349 and $1.483 \times \text{MAD}$.

The use of the absolute value in place of squaring also has an application in linear regression. We replace least squares with

$$(\hat{a}, \hat{b}) = \underset{(a,b)}{\arg \min} \sum_{i=1}^{n} |y_i - a - b(x_i - \bar{x})|.$$

The **R** function `l1fit` in the package `L1pack` performs the optimization. A simple thought experiment shows how effective this criterion can be. Suppose $n - 1$ pairs of the data are fit perfectly by a straight line. Then if the one point, (x_i, y_i), has a large value of y_i, least squares is strongly influenced, since we know the fitted line must go through the point $(\bar{x}, \bar{y})$. The L_1 fit, on the other hand, goes through the $n - 1$ points exactly; see Figure 9.9 for two examples. In each example, the L_1 linear regression line goes through the points on the line $y = 1 + 1.5x$.

At an advanced level, there are two general approaches to introducing robustness. The first is to modify the likelihood equations so that outliers are "downweighted." Huber (1972) gives a nice review in his Wald Lecture. An alternative approach is to replace MLEs with a minimum distance estimator. Here, we attempt to find the parameter vector, θ, that minimizes an estimate of the integrated squared distance between the parametric model and the unknown true density, $g(x)$. This is similar to our histogram bin width selection criterion in Equation (9.4). Scott (2001) introduces a fully data-based criterion called L_2E that achieves this goal by minimizing

$$\hat{\theta} = \underset{\theta}{\arg \min} \left[\int f_\theta(x)^2 \, dx - \frac{2}{n} \sum_{i=1}^{n} f_\theta(x_i) \right] \quad \text{or}$$

$$(\hat{\mu}, \hat{\sigma}) = \underset{(\mu,\sigma)}{\arg \min} \left[\frac{1}{2\sqrt{\pi}\sigma} - \frac{2}{n} \sum_{i=1}^{n} \phi(x_i \mid \mu, \sigma) \right] \quad \text{for normal data.} \qquad (9.16)$$

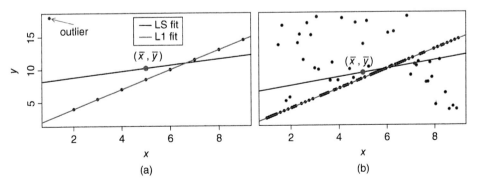

Figure 9.9 (a) $n = 9$ points satisfying $y = 1 + 1.5x$ with one outlier at $x = 1$; (b) $n = 101$ with 40 randomly selected outliers (with negative slope).

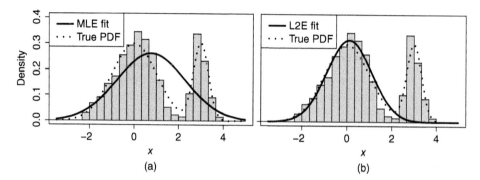

Figure 9.10 (a) MLE $N(\bar{x}, s^2)$ and (b) L_2E normal fits to a random sample of 400 points from the mixture $0.75 \times N(0, 1) + 0.25 \times N(3, 1/3^2)$; see text.

An example of a normal fit by MLE and L_2E applied to a mixture of two normal densities is displayed in Figure 9.10. The fitted model is, of course, incorrect. However, the difference in the fitted models strongly hints at the benefits of robust modeling. The MLE fit uses $\bar{x}$ and $\sqrt{s^2}$ as usual. The L_2E model fitted is actually a three-parameter normal model, $w \cdot N(\mu, \sigma)$, where w is a scaling factor. The L_2E criterion in Equation (9.16) is modified by introducing the factors w^2 and w in the two terms, respectively. See also Basu et al. (1998) for a more general family of divergence criteria, which includes L_2E as a special case.

9.9 Conclusions

This concludes our selection of research examples and advanced topics. The fundamental material in the first eight chapters allows us to understand how those concepts extend to all advanced material in an intuitive fashion. I hope this concise textbook encourages and facilitates your future studies. I wish you continued success.

Appendices

A   Notation Used in This Book

Notation	Remarks
I_A	0/1 indicator function for occurrence of event A
log	natural logarithm (ln)
$\log_{10}$	common logarithm
w.l.o.g.	without loss of generality
$\text{sgn}(x)$	sign function: 1 if $x > 0$; -1 if $x < 0$; and 0 if $x = 0$
$\forall i$	for all values of i
$\sum_{i=1}^{n} x_i$	summation $x_1 + x_2 + \cdots x_n$
$\prod_{i=1}^{n} x_i$	product $x_1 x_2 \cdots x_n$
$\mathbb{R}, \mathbb{R}^1$	the real number line
$\mathbb{R}^d$	Euclidean space of dimension d
A^c	the complement of event/set A
$A \cup B$	the union of events A and B
$A \uplus B$	the disjoint union of events A and B
$A \cap B$	the intersection of events A and B
AB	a notational shorthand for $A \cap B$
$\emptyset$	the null set
Ω	sample space or sure event; note $\Omega = \emptyset^c$
$\mathcal{F}$	list of all sets in an experiment (an Algebra of sets)
$n!$	n factorial; $n! = n \times (n-1) \times \cdots \times 2 \times 1$. Note: $0! = 1$.
$^n P_r$	permutation function, selecting r from n objects w/o replacement
$^n C_r = \binom{n}{r}$	combination function, like permutations without order

Statistics: A Concise Mathematical Introduction for Students, Scientists, and Engineers, First Edition. David W. Scott.
© 2020 John Wiley & Sons Ltd. Published 2020 by John Wiley & Sons Ltd.

Notation	Remarks
$F(x)$	cumulative distribution function (CDF)
$p(x)$	probability mass function (PMF)
$f(x)$	probability density function (PDF)
$X \sim f$	the random variable X has density f
r.v.	random variable
i.i.d.	independent and identically distributed (r.v.s)
x	bold notation for the $n \times 1$ vector with elements $x_1, x_2, \ldots, x_n$
$\mathbf{x}^T$	$1 \times n$ transpose of the $n \times 1$ vector **x**
$d\mathbf{x}$	bold notation for product $dx_1\ dx_2\ \cdots\ dx_n$
$E(X) = \mu$	expectation of X
$\mathrm{var}(X) = \sigma^2$	variance of X
$\mathrm{cov}(X, Y)$	covariance of X and Y (γ_{XY})
$\mathrm{cor}(X, Y)$	correlation of X and Y (ρ_{XY})
MGF $M_X(t)$	moment generating function $E[\exp(tX)]$
$\Phi(x)$	CDF of the standard normal $N(0, 1)$ density
$\phi(x)$	PDF of the standard normal $N(0, 1)$ density
$\Phi(x|\mu, \sigma^2)$	CDF of the normal $N(\mu, \sigma^2)$ density
$\phi(x|\mu, \sigma^2)$	PDF of the normal $N(\mu, \sigma^2)$ density
$\chi^2(p)$	PDF of the chi-squared density with p degrees of freedom; also χ^2_p
$P(\cdot)$	probability of an event (or $\mathrm{Pr}\,(\cdot)$)
$\mathrm{Binom}(n, p)$	binomial PDF, n trials with probability p
$\mathrm{Unif}(a, b)$	uniform PDF on the open or closed interval
$\mathrm{Pois}(m)$	Poisson PDF with parameter m
$\Gamma(x)$	Gamma function, generalized factorial function $\Gamma(n) = (n - 1)!$
$\mathrm{Beta}(\mu, v)$	Beta PDF
t_p	Students' T PDF
$F_{r,s}$	Snedecor's F PDF
$a_n = O(b_n)$	$\Longleftrightarrow a_n/b_n \to c$ as $n \to \infty$ ("big O")
$a_n = o(b_n)$	$\Longleftrightarrow a_n/b_n \to 0$ as $n \to \infty$ ("little o")
$X_{(i)}$	ith order statistic, $X_{(1)} \le X_{(2)} \le \cdots \le X_{(n)}$
x_p	pth percentile in a sample of size n, roughly, $x_{(np)}$
Median	$x_{0.5}$
IQR	interquartile range, $x_{0.75} - x_{0.25}$
Binomial theoremm	$(x + y)^n = \sum_{i=0}^{n} \binom{n}{i} x^i\, y^{n-i}$
Geometric	$\sum_{k=0}^{n} r^k = (1 - r^{n-1})/(1 - r)$; $\sum_{k=1}^{n} r^k = (r - r^{n-1})/(1 - r)$
series	$\sum_{k=0}^{\infty} r^k = 1/(1 - r)$; $\sum_{k=m}^{\infty} r^k = r^m/(1 - r)$ if $|r| < 1$
Exponential	$e^x = \sum_{i=0}^{\infty} x^i/i!$

B   Common Distributions

Let $q = 1 - p$. The range of t around 0 where the MGF exists is not given.

Binomial:
$$p_X(x) = \binom{n}{x} p^x q^{n-x} \quad x = 0, 1, \ldots, n$$

Binom(n, p)
$$M_X(t) = (pe^t + q)^n \quad \mu = np \quad \sigma^2 = npq$$

Geometric:
$$p_X(x) = pq^{x-1} \quad x = 1, 2, \ldots$$

Geom(p)
$$M_X(t) = \frac{pe^t}{1 - qe^t} \quad \mu = \frac{1}{p} \quad \sigma^2 = \frac{1-p}{p}$$

Negative binomial:
$$p_X(x) = \binom{r+x-1}{x} p^r q^x \quad x = 0, 1, 2, \ldots$$

NegBinom(r, p)
$$M_X(t) = \left(\frac{p}{1 - qe^t}\right)^r \quad \mu = \frac{rq}{p} \quad \sigma^2 = \frac{rq}{p^2}$$

Poisson:
$$p_X(x) = \frac{e^{-m}m^x}{x!} \quad x = 0, 1, 2, \ldots$$

Pois(m)
$$M_X(t) = e^{m(e^t - 1)} \quad \mu = m \quad \sigma^2 = m$$

Multinomial:
$$p_X(\mathbf{x}) = \binom{n}{x_1, x_2, \ldots, x_K} p_1^{x_1} p_2^{x_2} \cdots p_K^{x_K}, \quad \sum_{k=1}^{K} x_k = n$$

MultiNom$(n, \mathbf{p})$
$$\mu_k = np_k \quad \sigma_k^2 = np_k(1 - p_k) \quad \text{cov}(X_k, X_\ell) = -np_k p_\ell$$

Uniform:
$$f(x) = \frac{1}{b-a} I_{(a,b)}(x)$$

Unif(a, b)
$$M_X(t) = \frac{e^{bt} - e^{at}}{(b-a)t} \quad \mu = \frac{a+b}{2} \quad \sigma^2 = \frac{1}{12}(b-a)^2$$

Exponential:
$$f(x) = \frac{1}{\beta} e^{-x/\beta} \quad x, \beta > 0$$

exp(β)
$$M_X(t) = \frac{1}{1 - \beta t} \quad \mu = \beta \quad \sigma^2 = \beta^2$$

Normal:
$$f(x) = \frac{1}{\sqrt{2\pi}\sigma} e^{-\frac{(x-\mu)^2}{2\sigma^2}} \quad x, \mu \in \mathbb{R}^1 \ \sigma > 0$$

$N(\mu, \sigma^2)$
$$M_X(t) = e^{\mu t + \frac{1}{2}\sigma^2 t^2} \quad \mu = \mu \quad \sigma^2 = \sigma^2$$

Chi-squared:
$$f(x) = \frac{1}{\Gamma(p/2)2^{p/2}} x^{\frac{p}{2}-1} e^{-x/2} \quad x \geq 0 \ p = 1, 2, \ldots$$

$\chi^2(p)$
$$M_X(t) = \left(\frac{1}{1-2t}\right)^{p/2} \quad \mu = p \quad \sigma^2 = 2p$$

Students' T:
$$\frac{\Gamma((p+1)/2)}{\Gamma(p/2)}(\pi p)^{-1/2}\left(1 + \frac{x^2}{p}\right)^{-(p+1)/2} \quad x \in \mathbb{R}^1 \ p = 1, 2, \ldots$$

t_p
$$\mu = 0 \ (p > 1) \quad \sigma^2 = \frac{p}{p-2} \ (p > 2)$$

Snedecor's F:	$\dfrac{\Gamma((r+s)/2)}{\Gamma(r/2)\Gamma(s/2)}\left(\dfrac{r}{s}\right)^{r/2}\dfrac{x^{(r-2)/2}}{\left(1+\dfrac{r}{s}x\right)^{(r+s)/2}}$ $\quad x>0 \quad r,s=1,2,\dots$
$F_{r,s}$	$\mu = \dfrac{s}{s-2}\;(s>2)\quad \sigma^2 = 2\left(\dfrac{s}{s-2}\right)^2\dfrac{r+s-2}{r(s-4)}\;(s>4)$
Beta(α,β)	$\dfrac{\Gamma(\alpha+\beta)}{\Gamma(\alpha)\Gamma(\beta)}x^{\alpha-1}(1-x)^{\beta-1}\quad x\in(0,1)\,,\;\alpha,\beta>0$
	$\mu = \dfrac{\alpha}{\alpha+\beta}\quad \sigma^2 = \dfrac{\alpha\beta}{(\alpha+\beta)^2(\alpha+\beta+1)}\quad \text{mode} = \dfrac{\alpha-1}{\alpha+\beta-1}\;\;\alpha,\beta>1$
Gamma(α,β)	$\dfrac{1}{\Gamma(\alpha)\beta^\alpha}x^{\alpha-1}e^{-x/\beta}\,,\quad x\ge 0,\;\alpha,\beta>0$
	$M_X(t) = \left(\dfrac{1}{1-\beta t}\right)^\alpha\quad \mu=\alpha\beta\quad \sigma^2=\alpha\beta^2$
Cauchy	$\dfrac{1}{\pi(1+x^2)}\;x\in\mathbb{R}^1$ the MGF and moments do not exist

C   Using R and Mathematica For This Text

C.1   R Language – The Very Basics

1. **R** is a functional language, with arguments inside parentheses ()
2. Download **R** from https://cran.r-project.org/; see **R** Core Team (2018)
3. Type help.start() to bring up an interactive search window
4. We will use **R** in place of probability tables and for plotting
5. We will focus on basic **R**, not **R** Studio and other advanced options/libraries
6. Most **R** functions have optional arguments with default values
7. We end with a description of how to write your own **R** function

```
Basic Arithmetic, built-in Functions, and Graphics (the R prompt is >)

> x=1+3; y=sin(x); z=exp(x)      separate multiple inline commands with ;
> xyz = c(x,y,z)                 make a vector by concatenation

> xr=runif(100); yr=runif(100)   100 random points on unit square
> plot(xr,yr); lim=c(0,1)                basic plot of scatter diagram
> plot(xr,yr,pch=16,col=2,xlim=lim,ylim=lim,main="Example") # better?

> x = seq(0,1,0.01)       vector of 101 equally spaced points (0:100)/100
> y = sin(x)              y is also a vector with same length as x
> plot(x,y,type="l")      line plot rather than points (default)
> ls()                    unix-like command listing objects in file .RData

Probability Related Functions

For any built-in prob mass or density function, there are 4 functions.
The Uniform is unif, standard Normal is norm, the Binomial is binom, etc.
> fx = dnorm(x)           the pdf at x (x can be a vector)
```

```
> Fx = pnorm(x)            the cdf at x; Prob(X≤x)
> q = qnorm(p)             quantile: P(X≤q)=p
> x = rnorm(100)           generate 100 N(0,1) samples
> set.seed(123)            set random number seed, so can repeat the sequence
```

Looping and Control

```
> n=100; nrep=1000; ans=rep(0,nrep)        # create a new vector
> for(k in 1:nrep) {ans[k]=mean(rnorm(n)} # save xbar in kth place
> hist(ans,40,col=2,main="dist_of_xbar") # plot histogram of means
> if(runif(1)>0.5) {side="Heads"} else {side="Tails"}
```

Writing Your Own R **Function** to Visualize mean and std dev
Note: **Variables** created in functions are not saved on exit (good!)

```
> hw1 = function(n=100,nrep=1000,seed=123) { set.seed(seed)
>     ans=matrix(0,nrow=nrep,ncol=2)      # save simulation results
>     for( k in 1:nrep ) {      # simulation loop
>         z=rnorm(n); ans[k,1]=mean(z); ans[k,2]=sd(z)
>     }      # saved mean and standard deviation in kth row of ans
>     hist( ans[,1], 40, col=2 ); abline(v=0)    # 1st column of ans
>     browser()    # examine graph; type "c" to continue; "Q" to quit
>     hist( ans[,2], 40, col=2 ); abline(v=1)    # 2nd column of ans
>     return(list(ans=ans,n=n) }  # the function is now complete
> results = hw1()              # will it run the first time?
```

Making It Easy to **Do** Your Homework: Type Into the **File** hw1.txt
If you include the command results=hw1() at the end, it will execute below

```
> source("hw1.txt") # read the commands into R and debug if necessary
> dev.copy2pdf("fig1.pdf") # copy figure to a pdf file to submit
> print(results$ans); print(results$n)      # access items in the list
> q()            # quit R (save to .RData option given)
```

C.2 Mathematica – The Basics

1. Mathematica can perform both symbolic and numerical calculations; see Wolfram Research Inc. (2018)
2. Under the Help menu, select Wolfram Documentation for syntax/help
3. Mathematica objects/functions all begin with a capital letter
4. While **R** uses () for functions, Mathematica uses [ ], eg. Exp[x]
5. Type shift+return to execute/enter a Mathematica command
6. You can refer to a previous result by %5 (i.e. fifth equation)
7. Everything in Mathematica is a list, or a list of lists

Basic Arithmetic and Functions:

```
1+3                    4
y = 1+3                4 ( assignment operator = )
y == 5              False ( logical operator ==    type 2 equal signs )
y^2                   16
```

Sin[x]   x symbolic
Exp[x]

Simple Calculus Examples:

Integrate[**Sin**[x], x]   −**Cos**[x]   indefinite integral
Integrate[**Exp**[x], {x,1,2}]   e(e−1)   definite integral
Integrate[**Exp**[−a x], {x,0,**Infinity**}]   1/a (conditional on a>0)
Integrate[**Exp**[−a x], {x,0,**Infinity**}, **Assumptions** −> a>0]
fx = a **Exp**[−a x]   negative exponential PDF
Integrate[fx, {x,0,**Infinity**}   1   area under the PDF
Integrate[x fx, {x,0,**Infinity**}   1/a   the mean

mgf = **Integrate**[**Exp**[t x] fx, {x,0,**Infinity**}]   a/(a−t) the MGF
mgf = **Integrate**[**Exp**[t x] fx, {x,0,**Infinity**}, **Assumptions** −> a>t]

D[mgf, {t,1}] /. t −> 0   **Derivative** gives 1st noncentral moment 1/a
D[mgf, {t,2}] /. t −> 0   2nd deriv gives 2nd noncentral 2/a**2

ans = **D**[mgf, {t,2}]
Limit[ans, t −> 0]   long version of previous

Other Useful Functions for Computing and Simplifying:
Sum[1/x^2, {x,1,**Infinity**}]   **Pi**^2/6
Sum[a^k, {k,1,**Infinity**}]   a / (1−a)

Factor[%]   % refers to the previous equation
Expand[%3]   %3 refers to the previous equation number 3
TrigExpand[%]

Graphics

px = 2 * (x−1) * 2^(−x)   a probability mass function for x>3
Sum[px, {x,4,**Infinity**}]   sums to 1
Sum[x px, {x,4,**Infinity**}]   mean is 11/2 (var=13/4 sd=1.803)

DiscretePlot[px, {x,4,12}]   PMF appropriate
Plot[px, {x,4,12}]   Continuous curve (not appropriate)

Matrices and **Jacobian** /* **Box**−Muller Xform (u1,u2) −−> (z1,z2) */

z1 = **Sqrt**[−2**Log**[u1]] **Cos**[2**Pi** u2]
z2 = **Sqrt**[−2**Log**[u1]] **Sin**[2**Pi** u2]
u1 = **Exp**[−(z1^2+z2^2)/2]   /* **Box**−Muller Transformation */
u2 = **ArcTan**[z2/z1] / (2 **Pi**)   /* From 2−**D** Uniform to 2−**D Normal** */
jacob = { {**D**[u1,z1],**D**[u2,z1]}, {**D**[u1,z2],**D**[u2,z2]} }
Factor[**Det**[jacob]]   /* Bivariate Standard **Normal** PDF */

Bibliography

A. Basu, I. R. Harris, N. L. Hjort, and M.C Jones. Robust and efficient estimation by minimising a density power divergence. *Biometrika*, 85:549–559, 1998.

G. E. P. Box and D. R. Cox. An analysis of transformations. *Journal of the Royal Statistical Society, Series B*, 26:211–243, 1964.

G. E. P. Box and M. E. Muller. A note on the generation of random normal deviates. *The Annals of Mathematical Statistics*, 29:610–611, 1958.

G. Casella and R. L. Berger. *Statistical Inference, 2nd Edition*. Duxbury, Pacific Grove, CA, 2002.

T. Colton. *Statistics in Medicine*. Little, Brown and Company, Boston, MA, 1974.

J. L. Devore and N. R. Farnum. *Applied Statistics for Engineers and Scientists, 2nd Edition*. Duxbury Press, Pacific Grove, CA, 1999.

R. A. Fisher. On the mathematical foundations of theoretical statistics. *Philosophical Trans. Royal Society London (A)*, 222:309–368, 1922.

R. A. Fisher. *Statistical Methods for Research Workers*. Oliver and Boyd, Edinburgh, 1936.

D. Freedman and P. Diaconis. On the histogram as a density estimator: l_2 theory. *Zeitschrift für Wahrscheinlichkeitstheorie und Verwandte Gebiete*, 57:453–476, 1981.

M. Friendly. *HistData: Data sets from the history of statistics and data visualization*. R Foundation for Statistical Computing, Vienna, Austria, 2018. URL https://CRAN.R-project.org/package=HistData.

F. Galton. Regression towards mediocrity in hereditary stature. *Journal of the Anthropological Institute of Great Britain and Ireland*, 15:246–263, 1886.

J. Graunt. *Natural and Political Observations Made upon the Bills of Mortality*. Martyn, London, 1662.

P. J. Huber. The 1972 Wald lecture robust statistics: A review. *The Annals of Mathematical Statistics*, 43:1041–1067, 1972.

JMP®. *Version 14.3.0*. SAS Institute Inc., Cary, NC, 1989–2019. URL https://www.jmp.com/.

J. Neyman and E. S. Pearson. On the problem of the most efficient tests of statistical hypotheses. *Philosophical Transactions of the Royal Society of London. Series A*, 231:289–337, 1933.

R Core Team. *R: A Language and Environment for Statistical Computing*. **R** Foundation for Statistical Computing, Vienna, Austria, 2018. URL https://www.R-project.org/.

M. Rosenblatt. Remarks on some nonparametric estimates of a density function. *The Annals of Mathematical Statistics*, 1:832–837, 1956.

D. W. Scott. On optimal and data-based histograms. *Biometrika*, 66: 605–610, 1979.

Statistics: A Concise Mathematical Introduction for Students, Scientists, and Engineers, First Edition. David W. Scott.
© 2020 John Wiley & Sons Ltd. Published 2020 by John Wiley & Sons Ltd.

D. W. Scott. Parametric statistical modeling by minimum integrated square error. *Technometrics*, 43:274–285, 2001.

D. W. Scott. *Multivariate Density Estimation: Theory, Practice, and Visualization, 2nd edition.* John Wiley & Sons, New York, 2015.

S. S. Stevens. On the theory of scales of measurement. *Science*, 103: 677–680, 1946.

A. Stuart and J.K. Ord. *Kendall's Advanced Theory of Statistics, Volume 1, Distribution Theory 5th Edition.* Oxford University Press, 1987.

"Student". The probable error of a mean. *Biometrika*, 6:302–310, 1908.

H. A. Sturges. The choice of a class interval. *Journal of the American Statistical Association*, 21:65–66, 1926.

N. Taleb. *The Black Swan*. Random House, New York City, 2007.

I. M. Thompson, D. P. Ankerst, C. Chi, P. J. Goodman, C. M. Tangen, M. S. Lucia, Z. Feng, H. L. Parnes, and C. A. Coltman. Assessing prostate cancer risk: Results from the prostate cancer prevention trial. *Journal of the National Cancer Institute*, 98:529–534, 2006.

J W. Tukey. *Exploratory Data Analysis*. Addison-Wesley, Reading, MA, 1977.

Wolfram Research Inc. *Mathematica 12.0*, 2018. URL http://www.wolfram.com/.

Index

Statistics: A Concise Mathematical Introduction for Students, Scientists, and Engineers, First Edition. David W. Scott.
© 2020 John Wiley & Sons Ltd. Published 2020 by John Wiley & Sons Ltd.